# FASHION FIBERS
DESIGNING FOR SUSTAINABILITY

**Fairchild Books**
An imprint of Bloomsbury Publishing Inc

1385 Broadway
New York
NY 10018
USA

50 Bedford Square
London
WC1B 3DP
UK

**www.bloomsbury.com**

**FAIRCHILD BOOKS, BLOOMSBURY and the Diana logo are trademarks of Bloomsbury Publishing Plc**

First published in 2017



**Library of Congress Cataloging-in-Publication Data**
Names: Gullingsrud, Annie, author.
Title: Fashion fibers : designing for sustainability / Annie Gullingsrud.
Description: New York, NY, USA : Fairchild Books, an imprint of Bloomsbury Publishing Inc., 2016. | Includes bibliographical references and index.
Identifiers: LCCN 2016022589 | ISBN 9781501306648 (pbk. : alk. paper)
Subjects: LCSH: Textile fibers. | Textile design. | Sustainable engineering.
Classification: LCC TS1540 .G85 2016 | DDC 677/.022--dc23
LC record available at https://lccn.loc.gov/2016022589

ISBN: PB: 978-1-5013-0664-8
ePDF: 978-1-5013-0665-5

Cover design: Liron Gilenberg / www.ironicitalics.com
Cover image © Getty Images

Typeset by Lachina
Printed and bound in China

# FASHION FIBERS
# DESIGNING FOR SUSTAINABILITY

ANNIE GULLINGSRUD

Foreword by Lynda Grose
Illustrations by Amy Williams

**FAIRCHILD BOOKS**
AN IMPRINT OF BLOOMSBURY PUBLISHING INC.

B L O O M S B U R Y
NEW YORK • LONDON • OXFORD • NEW DELHI • SYDNEY

## Contents


vi Foreword by Lynda Grose
viii About the Author
ix Preface
xi Acknowledgments
xiii Introduction

2 Part 1

NATURAL FIBERS

5 1 Cotton
19 2 Flax
29 3 Bamboo Linen
39 4 Hemp
49 5 Jute
59 6 Wool
71 7 Silk
79 8 Leather
89 9 Alpaca
97 FUTURE FIBERS: NATURAL FIBERS
97 Piñatex™
98 CRAiLAR
99 Paper No. 9™

103 Part 2

MANUFACTURED FIBERS

105 10 Polyester
113 11 Nylon
121 12 Spandex
129 13 Acrylic
135 14 Imitation Leather
143 15 Polyethylene
149 16 Polypropylene (PP)
155 17 Rayon/Viscose Made from Wood
163 18 Rayon/Viscose Made from Bamboo
171 19 Lyocell
179 20 Modal
185 21 Azlon (From Soy)
193 FUTURE FIBERS: MANUFACTURED FIBERS
193 Polylactide
194 Polytrimethylene Terephtalate (Sorona® from DuPont)

199 Part 3

PROCESSING

201 22 Bleaching
209 23 Dyeing and Printing
225 24 Finishing
233 25 Garment Washing
245 FUTURE FIBERS: PROCESSING

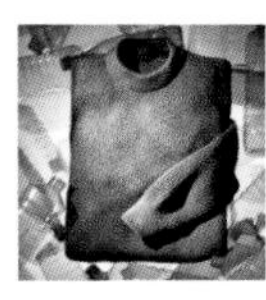

249 Part 4

RECYCLED/CIRCULAR TEXTILES

251 26 Recycled/Circular Textiles Technologies
259 FUTURE FIBERS: PROMOTING CIRCULAR TEXTILES
260 Evrnu
262 Re:newcell

265 Appendix A: Social and Cultural Sustainability
266 Appendix B: Consumer Care and Washing
268 Glossary
274 Bibliography
281 Notes
287 Index

# Foreword

by Lynda Grose

> *"Keep asking deeper questions about nature, about business and about the nature of business."*
> —Fritjof Capra

As I write this introduction, I can't help but reflect on the trajectory of which this book is part. A resource like this didn't exist when I first became aware of sustainability issues in fashion. Back then, in 1990, as an employee for the international clothing company ESPRIT, I worked on conventional fashion design—the look and style of garments—without considering ecological and social impacts. It was when the owner, Doug Tompkins, made a decision to use his company as a vehicle for communicating ecological issues and changing the way we conducted business that my sustainability education began. Recognizing the need for ecoliteracy throughout the organization, Tompkins engaged major environmentalists (David Brower, Dave Foreman, Randy Hayes, Paul Erlich, Frijof Capra, Earnest Callenbach, and many others) in a company-wide "Be Informed" lecture series, which employees were required to attend. Also initiated was the "eco-desk," where a full-time ecologist provided support and led a company-wide eco-audit.[1] Linking issues like wilderness preservation, limits to growth, and increasing population to designing garments was a stretch in those days. But the lectures and eco-audit provided me with a solid scientific and philosophical foundation and unprecedented conditions to be able to research, think through, and test practical solutions for a large corporation. This "research through practice" experience and the tacit knowledge that "working in the trenches" brings continue to inform the design for sustainability curriculum I teach at California College of the Arts, to this day. And it was in those classes that Annie Gullingsrud was inspired to choose a particular path for her own career, including identifying the need for and writing the book you now have in your hands.

Ours is a material world and materials are essential to sustainability ideas; materials visibly connect us to many of the big issues of our time: climate change, waste creation, and water poverty can all be traced back somehow to the use and processing of and demand for materials.[2] Yet few everyday people, even nowadays, are fully aware of the impacts of manufacturing fiber, cloth, and clothing. We find the technical complexity of textile processing bewildering. We feel less qualified than "the experts" and tend to leave the technical decisions and their implications for water courses, air quality, soil toxicity, and human and ecosystem health in the hands of textile scientists.

Sustainability in the fashion industry has to date therefore been an industry-led movement. And yet, for those companies that are diligently researching impacts of their production and devising mitigation strategies, communicating their actions to consumers is often too complicated. As a result, information provided to consumers tends to be oversimplified, hindering the knowledge gap, perpetuating the lack of deep understanding, and marginalizing the role of designers and wearers in developing solutions.

This book aims to disrupt this *cycle of not knowing* and to spark agency in its readers. Written in accessible language and arranged in an easy-to-understand fiber life cycle format, each chapter leads the reader through a range of issues from harvesting and extraction of raw material to consumer care and disposal options. In doing so, Gullingsrud breaks down the "natural is good and synthetics are bad" stereotype and helps the reader to realize that *all* fibers have impacts in a concentrated and industrialized fashion system. This book also provides basic information that can be used to shape informed questions of processors and companies, leading to wider disclosures . . . and since designers and wearers both provide the market for industrial textile supply chains, we can be catalysts for new developments.

As one might expect from a designer, besides providing practical technical information, Gullingsrud also suggests creative responses relevant to the specific impacts of each fiber. She identifies points of impact not as restrictions but as points of intervention and rich sources for innovation. In short, the author presents an optimistic view of fashion and textile sustainability and equips designers and consumers with knowledge to take action.

Studies show that as human population and wealth per person increase, we tend to use more natural resources, transforming them into consumer goods for sale. The exponential growth in raw material use is also attributed to current business models, which demand that companies grow exponentially each year by selling more units. Sustainability requires that humans learn to live and conduct business within the limits of the natural resources that are available for *all* species. So beyond the impacts of individual fibers, this book also introduces alternatives to linear systems of production, sometimes referred to as take-make-waste. Closed-loop material systems enable the recycling of fibers at the end of a product's useful life into second and third generations, and even in perpetuity. This is a critical concept, because all impacts and resource depletion in particular are slowed when more fibers are drawn from an already used source versus a virgin source.

The late Ernest Callenbach noted, "Things that are included in the vocabulary of companies gain familiar reality. Things that are left out are ignored or even have their existence denied."[3] Using this book as a reference tool on a daily basis, then, starts to influence the usual terms we use to talk about fashion practice and to shift the way we think about the fashion system. The "ecology of fashion practice," the "metabolism of a company" re-situate our thinking in context with the planetary systems that support all commercial practice, including fashion. At the very least this book helps us to bring a fresh set of values to influence and unfreeze industrial norms.

Yet looped product systems are not solely enabled through material recycling. They are also present in social and cultural behavior. We all have garments we have kept for a long time, used differently than intended, adapted to our own bodies and tastes, gifted, or passed down to a friend or family member, for example. As the fashion research project *Local Wisdom*[4] reveals, these human systems—of exchange, repair, creation, sharing—are innately sustainable, yet they are seldom observed and often missed by companies because they fall outside the core commercial purpose of making things to sell as *the* means to generate revenue. Ecological benefits of *use behaviors* therefore remain uncommunicated, invisible, unacknowledged, and undervalued by companies and customers alike.

So as you use this book, be aware of the place you have in its trajectory. Use its contents and test them in your own "craft of practice" circumstances. Note when existing industrial systems and structures limit the application of its contents, identify these new points of intervention, and pursue them relentlessly. Above all, use this book well—that is to say, be informed by its content but also let its practical use re-inform and expand the book's premise. For this is one of many, many references that we will need to truly change the fashion system and commerce as a whole so that all can thrive within Earth's ecological boundaries.

Thank you for purchasing this book.

Lynda Grose
Designer and Educator
California College of the Arts, USA

*Lynda Grose has worked in the fashion industry for more than thirty years and has spent most of her career focused on sustainability issues. She co-founded ESPRIT* **eco***llection, the first ecological line of clothing marketed internationally by a global corporation, and has worked with designers, artisans, farmers, and corporate executives to further sustainability in the fashion sector. She sees design as a tool to help give form to a sustainable society and is passionate about new roles for design in achieving this end.*

# About the Author

**Annie (McCourt) Gullingsrud** has progressive experience in both marketing communications and fashion design and has worked as a sustainability consultant, writer, and designer.

After seven years of working at marketing and advertising agencies, Gullingsrud's hands started to itch. She wanted to sew and make things. She went back to school to study fashion design at California College of the Arts in San Francisco. The introduction to sustainable fashion design brought purpose, meaning, and richness to a field of study that she had begun to think was lacking direction. The introduction to Cradle to Cradle design principles and methodology brought optimism, joy, and the perfect solution. She had found her path.

Gullingsrud has studied natural dyeing and weaving with local artisans in Guatemala; worked as a fashion designer at a sustainably run factory in Madhya Pradesh, India; and developed a process of cutting and patterning that eliminates pattern-cut waste. She has written and designed a book about fibers used in the fashion industry for employees of for Gap Inc. and TEKO, Swedish Textile Association, to teach them how to design around and reduce impacts. Gullingsrud is currently the director of the Fashion Positive Initiative (www.fashionpositive.org), at the Cradle to Cradle Products Innovation Institute and guides fashion and textile brands, designers, and suppliers toward the creation of optimized materials for the circular economy.

# Preface

*Sustainability.* A word often used but not consistently defined. How do we determine sustainability? What does it mean? How do I trust claims of sustainability?

When I was a student studying fashion design and working on my sustainable design thesis as a senior, I found myself often confused by fabric selection. I wanted to use better quality fabrics, as defined by their environmental and social attributes. I noticed claims such as "sustainable," "eco-friendly," or "environmentally friendly" without any consistency for how these descriptions were used. I decided to do my own research. Once I looked deeper at the fabrics themselves, the fibers making up these fabrics, and their production processes, it seemed as if these claims were merely for marketing purposes. In most cases I found that while one aspect of these fabrics could be considered environmentally friendly, there were often other "negative" aspects that seemed to outweigh the positive ones.

Sustainability means that materials are grown, made, and "disposed of" in a safe and beneficial way—to all living things and the planet. I've found that so few fibers on their own can actually be considered "sustainable" (I can count only one).

Sustainability is an action. Sustainability is not only defined by the fibers we select, but also by how we process and use them. It's up to us to influence the potential for a fiber, and the resulting garment, to be considered sustainable.

For this reason, *Fashion Fibers: Designing for Sustainability* sees the sustainability potential in most fibers used in the fashion industry. This book encourages an all-inclusive approach that believes that all fibers, with the right decision making, a little bit of work, creativity, and especially innovation, have the potential to be designed and produced sustainably.

The recommendations are not dictated. *Fashion Fibers: Designing for Sustainability* is an unbiased approach to fiber sustainability that evaluates the checks and balances of each fiber and process, giving *you* an opportunity to decide for yourself what is intuitively important to your values and/or your company.

Each chapter in *Fashion Fibers: Designing for Sustainability* is organized in the same way, offering an easily accessible reference tool. *Fashion Fibers: Designing for Sustainability* functions as both an educational text for learning about sustainable fiber options and a reference guide for making fiber selection and decision making throughout the design and production stages.

*Fashion Fibers: Designing for Sustainability* investigates six main areas of potential impact in fiber **cultivation**, production, and processing, including chemical use, water (use and effluent), fair labor, energy use, consumer use/washing, and biodegradability/recyclability. This book provides knowledge as to where these impacts exist for a fiber and its resulting garment, and offers suggestions on how to design around, reduce, and capitalize on the positive impacts.

## Organization of the Text

*Fashion Fibers: Designing for Sustainability* consists of four parts: Part 1 presents natural fibers and Part 2 presents manufactured/synthetic fibers. Part 3 presents fiber/garment processing most often used in the fashion/textile industry, such as dyeing and finishing, and Part 4 presents promoting circular textiles. A Future Fibers section closes each part of the text, showcasing emerging fiber technologies and innovations—some still in the pilot or research stage and unable to be categorized in traditional fiber categories.

Each chapter will follow a consistent outline throughout the text:

- **a. Benefits:** Discusses environmental benefits associated with cultivation/production, processing, and intrinsic fiber characteristics.
- **b. Potential impacts:** Discusses environmental impacts with cultivation/production, processing, consumer care, and washing.
- **c. How to optimize sustainability benefits:** This section is presented in a table format and makes suggestions on how to guide your decision making to capitalize on sustainability benefits that each fiber intrinsically has.

d. **Availability:** Industry availability, supplier references for sustainable materials.
e. **Fashion applications:** Suitable applications for garments, such as "blouse," "denim," or "outerwear."
f. **Marketing opportunities:** List of opportunities to market fiber and what can and cannot be claimed. These claims should also be substantiated by research.
g. **Innovation exercises:** An interactive section that gives you the opportunity to practice what you've learned through a series of exercises that will inspire you to design for sustainability.

Sustainability can be a guiding force for creativity and innovation in everything that you do. It all starts with knowing where the opportunities lie. *Fashion Fibers: Designing for Sustainability* will help you do just that.

### Instructor Resources

*Fashion Fibers: Designing for Sustainability* provides an instructor's guide, test bank, and a PowerPoint presentation for instructors who would like additional support in teaching and introducing this text. To facilitate the integration of the online student resources, a *Learning with STUDIO* Student Registration Guide (PDF) and a *First Day of Class* PowerPoint presentation are also available. Instructor's Resources may be accessed via Bloomsbury Fashion Central (www.BloomsburyFashionCentral.com).

### STUDIO™ Fashion Fibers: Designing for Sustainability

Fairchild Books has a long history of excellence in textbook publishing for fashion education. Our new online STUDIOS are specially developed to complement this book with rich media ancillaries that students can adapt to their visual learning styles. *Fashion Fibers: Designing for Sustainability* STUDIO features online self-quizzes with scored results and personalized study tips, real-world case studies, and flash cards with terms and definitions to help students master concepts and improve grades. The STUDIO also contains a biodegradability appendix, which compiles information about designing for biodegradability.

STUDIO access cards are offered free with new book purchases and also sold separately through Bloomsbury Fashion Central (www.BloomsburyFashionCentral.com).

# Acknowledgments

I've been shaped by the world and people around me. I am so grateful that the world and people around me are so delightful!

To Lynda Grose: Without you and the sustainability education you gave me at California College of the Arts, I still might be looking for meaning in this world. You delivered me my passion; I will be forever finding ways to pass that on to others.

To Susan McMullen: Thank you for the countless hours you spent assisting me with the book's research. You are a delight to work with and it would give me great joy to get to work together again.

To Amy Williams: The education you gave me at California College of the Arts prepared me to bring my knowledge, passion, and enthusiasm out into the world. What a joy to get to work with you professionally—thank you for bringing the book to life through your illustrations.

To Fashion Positive and the Cradle to Cradle Products Innovation Institute: Thank you for the pathway that helps me walk positively and hopefully on a daily basis.

To Lewis Perkins: I am energized and inspired by the vision you bring. Thank you for supporting me. This book wouldn't have been written without your encouragement.

To my husband, Miles Gullingsrud: We were married a mere four weeks when I began writing this book. What could have been a potential strain on our relationship really turned into another opportunity for you to express your love and support. I love you.

To Amanda Breccia, Amy Butler, and Edie Weinberg at Fairchild Books: Thank you for your direction throughout the writing process and for always being available to provide guidance.

To the book's reviewers: Dr. Melinda K. Adams, University of the Incarnate Word; Sonja Andrew, University of Manchester, UK; Gail I. Baugh, San Francisco State University; Beverley Bothwell, University of Bedfordshire, UK; Leslie Davis Burns, Oregon State University; Huantian Cao, University of Delaware; Kelly Cobb, University of Delaware; Kim Hiller Connell, Kansas State University; Aleta Deyo, Mount Ida College; Melissa Halvorson, Marist College; Harriet McLeod, Kent State University; Amy Williams, California College of the Arts; Dilys Williams, London College of Fashion, UK; and Theresa Winge, Michigan State University: Thank you for your time and dedication to this book. Your feedback has improved the overall quality of the book and its focus.

Thank you to Gap Inc., Social & Environmental Responsibility, Kindley Walsh Lawlor, and Melissa Fifield for contributing to widely dispersing knowledge for the fashion industry by working with me to develop the foundation material for this book.

To the industry-expert reviewers (my colleagues and friends): You are brilliant people. Not only do I get to work with you, but I get to learn from you on a daily basis. Thank you for taking the time to review and provide invaluable feedback that puts this book in a better position to be used:

- Jay Bolus, President, Certification Services, McDonough Braungart Design Chemistry (MBDC)
- Lynda Grose, fashion designer, author, consultant, associate professor fashion and sustainability, California College of the Arts
- Bruce Nelson, Alpaca breeder and rancher, Ahh . . . Sweet Alpacas
- Tricia Carey, director of Global Business Development, Lenzing
- International Wool Textile Organization
- Australian Wool Innovation
- Tone Tobiasson, editor, nicefashion.org
- Kjersti Kviseth, life cycle designer, 2025design, Norway
- Nick Morley, director, Faering Ltd.
- Nicole Rycroft, executive director, Canopy
- Tara Sawatsky, corporate campaigner, Canopy

To the designer of the planet: I am constantly awed by your brilliant design. It is my hope that we can leverage this brilliance as much as possible, to work with it, be smarter, and not waste it.

Bloomsbury Publishing wishes to gratefully acknowledge and thank the editorial team involved in the publication of this book:

Senior Acquisitions Editor: Amanda Breccia

Development Editor: Amy Butler

Assistant Editor: Kiley Kudrna

Art Development Editor: Edie Weinberg

Photo Researcher: Rona Tuccillo

Illustrator: Amy Williams

In-House Designer: Eleanor Rose

Production Editor: Claire Cooper

Project Manager: Chris Black, Lachina

STUDIO and Instructor Resources: Leslie Davis Burns

# Introduction

The production of a fiber that makes up the resulting garment or fashion product can influence a wide range of negative environmental and social impacts. To give you a little perspective on how much influence, the Danish Fashion Institute claimed in 2013 that fashion is the world's second-most polluting industry, second only to oil. There are many common—yet unintended—side effects to fashion and textiles production. **Climate change**, resource depletion, impacts on human health, overuse of nonrenewable resources (as in oil in synthetics production), water and air pollution, excessive waste, and loss of **biodiversity** are just some of the impacts that the cultivation, processing, use, consumption, and disposal of fibers can create.

Since sustainability is an action, what's even more important are the design decisions you make from conception to end of use, because these design decisions will influence the overall sustainability of a garment. As a designer, you can have a significant amount of influence whether the clothes you design have a positive or negative impact.

## Product Life Cycle and Considerations

There are approximately fourteen stages in a product's life cycle. Each of those stages has the potential for impacts, both negative and positive. For example, in the cultivation of a material, depending on what it is, we want to circumvent resource depletion, **deforestation**, and desertification and want to instead promote renewability, good farming practices, fair treatment of animals, and biodiversity. In the production and processing of a fiber into a fabric and the resulting garment, we want to circumvent extreme water and energy use, minimize toxic chemicals, and reduce $CO_2$ **emissions**, excessive waste generation, untreated water discharge, and emissions from transportation. Instead, we want to promote closed-loop systems for water recovery and recycling, treated water discharge, renewable energy, chemicals that are verified to be safe for people and the planet, and innovative solutions for waste (whether at the design stage via cutting and patterning techniques, or at the post-industrial stage via recycling methods). At the end-of-use stage, we want to circumvent disposal to landfill and incineration and instead promote compostability and recyclability. You will learn ideas on exactly how to do this in the following chapters.

## Challenges

The main reason this is an entire book rather than just a list of "approved fibers" is that there is no single solution to sustainable fiber selection. Each fiber has different impacts at different points in its life cycle, and appropriate (and actionable) responses can be complex. Lack of transparency around manufacturing practices, traceability, a fragmented industry, lack of agreement on the

**STAGES OF A GARMENT'S LIFE CYCLE**

Each stage of a product's life cycle offers many opportunities to make choices that can turn a negative impact into a benign or positive influence on environmental and social conditions.

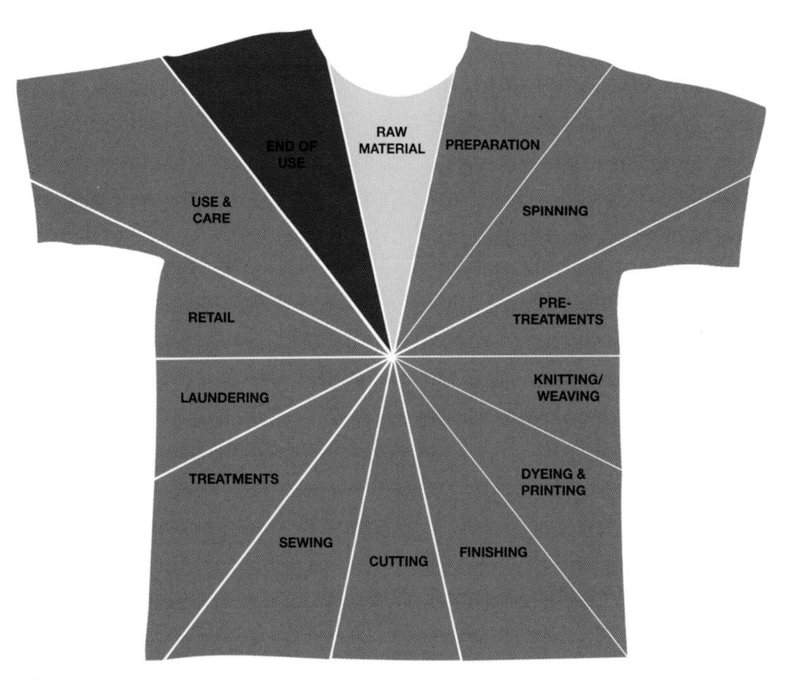


**Figure A**
Stages of a garment's life cycle offer negative impacts that can be circumvented by positive ones.

*Annie Gullingsrud*

means to measure and assess life cycle data, and cost all influence the decisions for sustainability we are able to make on a daily basis. Trade-offs are inevitable.

These challenges will lead you to creativity and innovation. In order to innovate around fiber and fabric sustainability, understanding the impacts of each fiber that you use is essential. This book is your resource to do just that. Knowing the options and processes currently available can lead to rapid adoption of alternatives, with multiple benefits.

## Sustainability and the Fashion Industry

In the past five years, there has been a significant growth in available sustainability efforts in the fashion industry. The Sustainable Apparel Coalition, an industry-wide coalition of brands working together to promote sustainable production, now includes Nike, Target, Walmart, Levi Strauss & Co., M&S, C&A, Gap Inc., Patagonia, Adidas, C&A, Uniqlo, H&M, Inditex (Zara), Lululemon, PUMA, REI, and many more.

Fashion Positive Initiative was launched in 2014 by the Cradle to Cradle Products Innovation Institute to make a positive impact on fashion through cradle-to-cradle design and production. Since then it has collaborated with brand partners such as Stella McCartney and Bionic to grow a Cradle to Cradle Certified materials library that is public and available to the industry.

Knowledge about sustainability of fiber and fabrics is becoming a requirement in the fashion industry. Use this book to arm yourself with facts, get educated, and get inspired about all of the creative possibilities that designing for sustainability can bring to you and your company.

*Alex Salinas*

## Part 1

# NATURAL FIBERS
## Overview

Part 1: Natural Fibers includes natural fibers that are divided into two main classifications: animal (protein) fibers and plant (cellulose) fibers.

- **Protein fibers** are derived from animals and include alpaca, leather, silk, and wool.
- **Cellulose fibers** are derived from plants, are products of agriculture, and are categorized as either bast fibers or seed fibers.
- Bast fibers are fibers collected from bast or skin of the plant and include fibers such as flax, jute, hemp, and bamboo (for linen).
- Cotton is a seed fiber and is grown as a boll around the seed of the cotton plant.

The following chapters highlight environmental and social benefits and impacts and will show you how to optimize the fiber's sustainability benefits to manifest innovation and effectiveness.

You will learn that each of these fibers inherently possesses natural sustainability attributes. The goal is to accentuate and encourage these natural attributes at every stage of fiber and fabric processing, and throughout the resulting product's life cycle. Fully considering these "sustainable fibers" means making specific decisions to ensure that sustainable practices are being carried out from cultivation to end of use.

**Part 1 Table 1 Fast-Growing Renewable Fibers That Can Be Harvested at Least Once a Year**

| Fiber | Length | Timing |
| --- | --- | --- |
| Flax | 1 meter[1] | 3–4 months[2] |
| Jute | 1–4 meters | 3–4 months[3] |
| Hemp | 4 meters | 3 months[4] |
| Bamboo (for linen or viscose) | 24 meters | 40 days[5] |

**Part 1 Table 2 Retting Process Comparison Chart[6]**

| Retting type | Description | Advantage | Impacts | Duration of retting |
| --- | --- | --- | --- | --- |
| Dew retting | Plant stems are cut or pulled out and left in the field to rot. | Returns nutrients back into the soil. | Reduced fiber strength; low and inconsistent quality; influence by weather; and product is contaminated with soil. | 2–3 weeks |
| Water retting | Plant stems are immersed in water (rivers, ponds, or tanks) and monitored frequently. | Produces fiber of greater uniformity and higher quality. | Extensive stench and pollution; high cost; low-grade fiber. Requires water treatment maintenance. | 7–14 days |
| Chemical retting | Boiling and applying chemicals. | More efficient and can produce clean and consistent long and smooth-surface bast fiber within a short period of time. | Unfavorable color; high processing cost. The wastewater is concentrated and rich in chemicals and **biological** matter and must be treated before its release; otherwise it can harm aquatic ecosystems, receiving water bodies, and drinking water sources. | 75 minutes–hour |

*li jingwang/iStock*

# 1

# Cotton

Cotton is a seed fiber and is grown as a boll around the seed of the cotton plant. Cotton represents 35 percent of the world's textiles and is the second largest fiber category worldwide.[1] China, India, and the United States account for two-thirds of global cotton production.[2] The high global demand for cheap cotton fiber encourages large-scale intensive production with a potential for significant ecological and social impacts.

Cotton can be grown in a variety of ways: conventional, genetically modified (GM), certified organic, transitional organic/organic in-conversion, and **Integrated Pest Management (IPM)**. Cotton can also be grown in natural colors, eliminating the need to use dyes.

### Conventional Cotton

Conventional cotton is grown with a seed that has not been genetically modified, and there has been no special attention given to types of chemicals used or not used to grow the cotton.

### Genetically Modified Cotton

Due to its high susceptibility to pests and pathogens, and high pesticide consumption, cotton has been a major focus for genetic modification. Cotton is one of the top three genetically modified crops in the world, along with corn and soy. Although some European countries have banned the use of GM seed entirely, other countries have continued to rely on it.[3] In the United States and India in 2013, sources state that 90 percent of cotton crops were genetically modified.[4,5] GM is a relatively new technology and its long-term effects are not yet fully understood. So, the pros and cons of GM technology are still hotly debated. For example, in humid growing regions such as the U.S. state of Georgia, where boll weevil is rampant, **GM (Bt) cotton**, the most successful variety of genetically modified cotton, has shown significant pesticide reductions. One study of the effect of the adoption of GM (Bt) cotton in China showed positive environmental impacts by reducing pesticide use, as well as increased yield and a reduction in negative health effects on farmers.

On the other hand, **Herbicide tolerant (Ht) GM cotton**, otherwise known as "**Roundup Ready**," a seed developed by Monsanto, has led to a sharp increase in herbicide use, which has in turn caused genetic resistance in weeds. Farmers report having to use more toxic herbicides to suppress "super weeds."

To add to this complex picture, proponents of GM cotton note that herbicide tolerant varieties have also led to reduced soil tillage, which results in less topsoil erosion.

GM cottonseed is more expensive than regular cotton seed because the companies that sell the seeds do not allow saving and replanting next season; new seeds must be purchased each season. GM cotton is available in large quantities through normal supply channels and is priced the same as conventional cotton.

### Certified Organic Cotton

Certified organic cotton disallows the use of GM seed and restricts or disallows the use of many synthetic agricultural chemicals. Organic farming aims to self-stabilize agro-ecosystems and uses crop rotation and biological means to control pests and pathogens. Studies show that organic farming builds organic matter in the soil; reduces soil, air, and water contamination; increases soil fertility and biodiversity; reduces human and wildlife health hazards; and reduces greenhouse gas emissions resulting from the production of synthetic fertilizers and pesticides.[6] In some regions, organic cotton farmers may gain more income, though this is by no means universal or guaranteed. Potentially lower yields, higher labor costs, and a market price determined by the commodity price index often mean organic cotton cultivation is a tenuous financial risk for the farmer.

Organic cotton represents less than 0.7 percent of global cotton grown worldwide and is therefore in short supply.[7] See Table 1.1 for the five top organic cotton-producing countries. The price of organic cotton tends to be

**Table 1.1 Top Five Countries for Global Organic Cotton Production (2011–2012)**[8]

| Country | Metric Tons | Fiber Production (% of total) |
|---|---|---|
| India | 103,004 | 74.20% |
| Turkey | 15,802 | 11.38% |
| China | 8,106 | 5.84% |
| Tanzania | 6,891 | 4.96% |
| United States | 1,580 | 1.14% |

higher than conventional due to the increased labor costs necessary for its cultivation. Organic fiber also requires special processing since it falls outside the usual scope of the commodity cotton supply chain. For example, track and trace mechanisms must be implemented in developing countries where bar code and bale identification systems are not in place. This requires additional labor and monitoring. Certified organic cotton must also remain segregated through ginning and processing into yarn and fabric, and this special handling adds additional costs.

Certification by a third party ensures that all of the proper documentation is in place and that the organic claims are valid. Exact cultivation requirements vary from region to region, since each bioregion has different pest pressures and resources, but there is international reciprocity among most certification agencies accredited by the **International Federation of Agriculture Movements (IFOAM)**, which maintains organic farming standards, organic accreditation, and certification.

Organic certification does not necessarily guarantee low water use, fair labor practices, or a fair price to the farmer and/or farmworkers.

### Transitional Organic Cotton/Organic In-Conversion

In order to be certified as organic, cotton fields must go through a two-year transition period in Europe, and a three-year transition period in the United States, where disallowed chemicals are not used. Transitional organic cotton uses non-GM seed and is grown in the same manner as organic but is still transitioning through this two- or three-year phase to certification. Once the land on which it is grown has completed the two- or three-year requirement, the fiber can be labeled as organic.

**Transitional cotton** creates a bridge for farmers to switch from conventional to organic cultivation, and it may be cheaper than organic since it isn't yet certified and doesn't command the same "value added" in the market. However, transitional organic cotton is not as readily available as conventional, and reciprocity in labeling from nation to nation has yet to be established. Currently, labeling for transitional cotton is only allowed in Europe, where it is called "organic in-conversion" and must be certified under the European organic standard. The United States does not allow labeling of transitional. The Organic Exchange in Europe has a transitional cotton task force and is currently lobbying the U.S. Department of Agriculture (USDA) to allow labeling of transitional cotton in the United States.[9]

**Figure 1.1**
Bale of cotton.
*fmajor/iStock*

**BOX 1.1**
**THE VARIETY OF WAYS COTTON CAN BE GROWN**

Cotton can be grown in a variety of ways—conventional, genetically modified, certified organic, transitional organic/organic in-conversion and Integrated Pest Management (IPM). Once picked, ginned, and baled, it is very difficult to visually determine how the cotton was grown.

## Integrated Pest Management Cotton

Integrated Pest Management is an approach to farming that focuses on long-term prevention of pests by integrating biological controls, habitat manipulations, and modification of cultural practices. Some of these strategies include using beneficial insects, "decoy planting" (providing alternative plants for pests to attack instead of the crop), and attracting predator animals to feed on the pests. Pesticides are used only after monitoring and established guidelines indicate that pests exceed acceptable levels.[10]

There are many different IPM cotton initiatives globally, and IPM can be interpreted and implemented in many ways. However, there are two notable IPM programs emerging that are based on clear and accessible data, interpretations, and strategies: Better Cotton Initiative and the Sustainable Cotton Project.

### Better Cotton Initiative (BCI)

The Better Cotton Initiative was established in 2005 and is active globally addressing environmental, social, and economic issues in an integrated program that reduces water and chemical use. BCI does not allow child labor or bonded/forced labor; incorporates better treatment of female workers and the proper handling of and training in the use of pesticides and fertilizers; and encourages the use and development of a number of techniques that will reduce water use, including installing drip irrigation, water scouting strategies, land leveling, and mulching to improve water retention. BCI engages and supports all cotton supply participants, from producers to retailers, and provides a central digital repository of BCI bales with unique bale identification codes. BCI allows the use of genetically modified seed. BCI is currently working with farmers in Brazil, India, Mali, Pakistan, and China, with active retail members such as H&M, IKEA, and Levis Strauss and Co.[11] BCI "units" can be identified in the supply chain but do not necessarily end up in the final garment.

### Sustainable Cotton Project's Cleaner Cotton™ Field Program[12]

Cleaner Cotton™ uses non-GM seed and is grown using biological means to control pests. The Sustainable Cotton Project (SCP) provides technical support to growers to implement Integrated Pest Management practices on their cotton crop. SCP field scouts conduct weekly monitoring on fields registered in the Cleaner Cotton™ program and provide growers with a written report on the condition of the crop, observations about pest and beneficial insect populations, beneficial insect habitats, and any areas of concern. Cleaner Cotton™ growers avoid the thirteen most toxic chemicals used in conventional cotton production and use biological means as a first resort if an infestation occurs.[13] If pests threaten to cause significant economic losses and all biological options have been exhausted, the growers are allowed to use sprays, but never the disallowed chemicals. Because farmers have this "safety net" option, SCP has found that it can persuade more farmers to convert more acres to biological farming practices. Due to the larger scale, Cleaner Cotton™ is a more effective tool for reducing chemical use than the niche of organic is in California at this time. In addition, though GM technology is often promoted as achieving increased fiber yields for farmers, Cleaner Cotton™ also maintains yields and provides a further incentive for growers to test biological systems of farming.

Cleaner Cotton™ is grown and ginned in California, and each bale is labeled with the Cleaner Cotton™ trademark. All U.S. cotton bales, including those that are Cleaner Cotton™, are tracked by the USDA PBI (permanent bale identification system), which traces each bale back to the farmer and field where it was grown. Cleaner Cotton™ can be shipped anywhere in the United States and/or around the world for spinning and processing into fabrics and garments. In fact, 95 percent of California's cotton harvest, including Cleaner Cotton™, is shipped overseas.

## Colored Cotton

Colored cotton is a naturally pigmented fiber that grows in a variety of colors, including purples, mauve, and more commonly, brown and green shades. It may be cultivated organically, conventionally, or as Cleaner Cotton™. Colored cotton reduces the overall volume of dyes (and associated pollution),

**BOX 1.2**
**INTEGRATED PEST MANAGEMENT COTTON**

Sustainable Cotton Project's Cleaner Cotton™ Program takes a collaborative, holistic approach to growing cotton. Sustainable Cotton Project field scouts conduct weekly monitoring on fields registered in the Cleaner Cotton™ program and provide growers with a written report on the condition of the crop, observations about pest and beneficial insect populations, beneficial insect habitats, and any areas of concern.

**Figure 1.2**
Chad Crivelli is a participating farmer in the Sustainable Cotton Project (SCP) since it began in 2001. Chad touts many benefits of SCP, including money savings from reduced chemical applications; alternative solutions for insect control; benefits of less chemical usage on the environment. Chad also attends grower meetings with SCP experts and finds it's helpful to compare notes to make the best decisions around pest control.
*jenny@sustainablecotton.org*

**Figure 1.3**
Natural-colored cotton pale olive cardigan, a collaboration between Honest by and Muriée.
*Alex Salinas*

**BOX 1.3 SUSTAINABLE BENEFIT TO USING COLORED COTTON**

Since colored cotton grows in a variety of shades, it can be used undyed. Using undyed naturally colored cotton reduces the overall volume of dyes (and associated energy and water use and pollution).

since it can be used undyed. Colored cotton is in short supply and is much more expensive than white organic fiber. In addition, the **staple** length of colored cotton fibers is significantly shorter than white cotton, and this can cause quality problems when used in 100 percent colored cotton yarns and in finer-count yarns; yarns may snap on power spinning frames, on weaving equipment, and when sewing knit fabrics. However, blending colored cotton with longer-staple white cotton and/or developing two-to-three-ply yarns can counter this potential quality problem. The brown and green shades of colored cotton tend to darken with washing and fade with exposure to light. Extensive testing should be conducted to ensure yarns and fabrics meet acceptable commercial quality.

# Sustainable Benefits

Cotton is a **renewable natural resource** and is readily available and inexpensive.

Cotton fiber is almost pure **cellulose**; it is soft and breathable and absorbs moisture readily, making cotton clothes particularly comfortable in hot weather.

The fiber's high tensile strength in soap solutions renders cotton garments easy to wash, and dry cleaning is not required.

GM (Bt) cotton, the most successful variety of genetically modified cotton, has shown significant pesticide reductions. Herbicide tolerant (Ht) varieties have also led to reduced soil tillage, which results in less topsoil erosion.

Organic farming uses crop rotation and biological means to control pests and pathogens. Studies show that organic farming builds organic matter in the soil; reduces soil, air, and water contamination; increases soil fertility and biodiversity; reduces human and wildlife health hazards; and reduces greenhouse gas emissions resulting from the production of synthetic fertilizers and pesticides.[14] In some regions, organic cotton farmers may gain more income, though this is by no means universal or guaranteed.

Transitional organic cotton uses non-GM seed, is grown in the same manner as organic, but does not yet have an organic certification. Transitional cotton creates a bridge for farmers to switch from conventional to organic cultivation, and it may be cheaper than organic since it isn't yet certified and doesn't command the same "value added" in the market.

Integrated Pest Management is a holistic approach to farming that focuses on long-term prevention of pests by integrating biological controls, habitat manipulations, and modification of cultural practices. Pesticides are used only after monitoring and established guidelines indicate that pests exceed acceptable levels.[15]

Colored cotton reduces the overall volume of dyes (and associated pollution), since it can be used undyed.

In 100 percent form, cotton fiber fabric is **biodegradable** and can be broken down naturally into simpler substances by microorganisms, light, air, or water. Absolute biodegradability depends on the dyes and trims used, as well as route of disposal.

By-products from cotton include cottonseed oil, cottonseed meal (a source of protein for livestock), and cotton seed hulls, which are mainly used to feed livestock and can also be used for petroleum refining and plastics manufacturing.[16]

# Potential Impacts

## Chemical Use in Cultivation

Conventional cotton cultivation uses some of the most toxic chemicals (pesticides, defoliants, and fertilizers) available for use in agriculture and accounts for 16 percent of insecticide applications worldwide, more than any other crop.[17] Many of these chemicals are used by farmers in developing countries where education, access to information, and understanding of the dangers posed by hazardous chemicals are lacking. Consequently, cotton farmers may experience acute pesticide poisonings, which can result in illness and death. The use of toxic chemicals can diminish soil fertility over time, and **irrigation** runoff can pollute regional water bodies.

## Water Use in Cultivation

Cotton generally requires irrigated water. Water management practices vary from growing region to growing region and are influenced by a number of factors, including the farming system used, water costs, local climate, etc. Water footprint information for cotton products is available in the public domain, but because different organizations use different parameters for their calculations, water footprint estimates for a particular product vary widely and can be somewhat unreliable. For example, one source claims that it takes 5,678 liters of water to produce enough cotton for a pair of jeans, whereas another source reports that it takes 1,892 liters.[18,19] As apparel companies in the industry collaborate on information and strategies for sustainability, footprint tools will become more normalized, consistent, and reliable.

Cotton is mainly grown in Mediterranean, desert, or near-desert climates, where freshwater is in short supply. Consequently, much of the global cotton crop is irrigated, and this can have extensive impacts on regional freshwater resources. For example, in Uzbekistan and Kazakhstan, cotton irrigation water diverted from rivers flowing into the Aral Sea (once the fourth largest inland sea in the world) has reduced its size significantly. This has impacted wildlife and has also had devastating social impacts on former fishing villages in the area.[20]

## Forced Labor

In countries where cotton is handpicked, fair treatment of the farmworkers may be of concern. In Uzbekistan, for example, one of the largest exporters of cotton in the world, more than one million students, teachers, and doctors are forced by the national government to work in the country's cotton fields for a two-month period each fall.[21] In recent years, over 130 apparel companies have pledged to boycott Uzbek cotton, but the effectiveness of that effort is unclear. According to reports in the Russian press, China and Bangladesh purchased 83 percent of the 2013 harvest.[22]

## Consumer Care/Washing

Cotton is often washed and tumble-dried at high temperatures and can require pressing. Studies on consumer garment washing habits, looking at a typical woman's short-sleeve cotton T-shirt in size small, indicate that washing is the single largest contributor to greenhouse gas emissions (constituting 48 percent of the total value) compared to other life cycle consideratons such as manufacturing, transportation, packaging, retail, and disposal.[23]

**BOX 1.4**
**COTTON CERTIFICATIONS**

**Fair Trade Cotton**

The Fair Trade program is active globally, primarily in developing nations, and secures a minimum fiber price for the farmer aiming to cover the average costs of sustainable production. In addition to this baseline price for their fiber, farmers also receive a Fair Trade premium, which allows them to invest in community projects, such as schools, roads, or health care facilities. Most Fair Trade cotton is neither organic nor IPM. However, Fair Trade standards encourage sustainable farming practices by restricting the use of certain agrochemicals. Fair Trade minimum prices for organic cotton are generally set 20 percent higher than the Fair Trade conventional minimum prices.[24]

**OEKO-TEX® Certified Cotton[25]**

OEKO-TEX® is an independent, third-party certifier that offers two certifications for textiles: OEKO-TEX® 100 (for products) and OEKO-TEX® 1000 (for production sites/factories). OEKO-TEX® 100 labels aim to ensure that products pose no risk to health. These products do not contain allergenic dyestuffs or dyestuffs that form carcinogenic aryl-amines, and several other banned chemicals. The certification process includes thorough testing for a long list of chemicals.

- Visit **Chapter 23: Dyeing and Printing** for more information about the impacts of dyeing.

**Global Organic Textile Standard (GOTS)**

**Global Organic Textile Standard** is a widely used standard for assessment of textile products that contain a minimum of 70 percent organic fibers. The GOTS standard has environmental requirements, as well as requirements for wastewater treatment plants and labor, that must be met along the entire organic textiles supply chain.

- Visit **Chapter 22: Bleaching** for information on the environmental impacts of bleaching.
- Visit **Chapter 23: Dyeing and Printing** for more information on impacts associated with dyeing.
- Visit **Chapter 26: Recycled/Circular Textiles Technologies** for more information about recyclability.
- Visit **Appendix B: Consumer Care and Washing** for more information.
- Visit **Appendix: Biodegradability in the Fashion Fibers STUDIO** for information about designing for biodegradability.

# Availability

All the above-mentioned alternatives to conventional cotton are now available in yarns and knitted and woven fabrics.

Organic cotton is available in a variety of fabrics, from 100 percent organic cotton to numerous blends with other fibers.

Due to high demand, the availability of organic cotton products is restricted. Forward projections and contracting are essential to secure supply for future seasons.

The first harvest of Better Cotton was in 2010–2011 and was provided by farmers in Brazil, India, Mali, and Pakistan. The year 2014 was the third harvest season for Better Cotton from China and the second harvest for Better Cotton in Turkey.[26]

Cleaner Cotton™ is now available in bales, roving, yarns, knit fabrics, and final products.[27] The Cleaner Cotton™ trademark may be used on all Cleaner Cotton™ content items as long as the fiber is tracked and verified through the processing system.

**Table 1.2 Optimize the Sustainability Benefits of Cotton**

| Opportunity | Benefit | Considerations |
|---|---|---|
| Promote the use of organic cotton. | • Organic farming uses crop rotation and biological means to control pests and pathogens. Builds organic matter in the soil; reduces soil, air, and water contamination; increases soil fertility and biodiversity; reduces human and wildlife health hazards; and reduces greenhouse gas emissions resulting from the production of synthetic fertilizers and pesticides.[26] In some regions, organic cotton farmers may gain more income, though this is by no means universal or guaranteed.<br>• Consumers are familiar with the term "organic." | • Higher price. Less availability. Organic certification does not necessarily guarantee low water use, fair labor practices, or a fair price to the farmer.<br>• Design high value items in organic, which can absorb higher fiber costs.<br>• Design organic into items that are highly processed, so the fiber cost is a smaller percentage of the total garment cost.<br>• Forecast styles and volumes and contract organic fiber in advance to ensure supply. |
| Promote the use of biological IPM cotton. | • Holistic approach to farming that focuses on long-term prevention of pests.<br>• Pesticides are used only after monitoring and established guidelines indicate that pests exceed acceptable levels.[28]<br>• More limited marketing opportunity than organic, but this is shifting with the expansion of BCI. | • Lower price than organic, only slightly more expensive than conventional. Readily available fiber, limited spinners and supply chain participants.<br>• Forecast styles and volumes and contract IPM fiber in advance to ensure supply.<br>• BCI is not generally noted in POS materials, but on company websites. |
| Promote Fair Trade cotton. | • Strong marketing opportunity since there is wide consumer awareness of "Fair Trade." | • Higher price. Less availability. Forecasting/ planning may be required. Most Fair Trade cotton is neither organic nor IPM, but Fair Trade standards encourage sustainable farming practices by restricting the use of certain agrochemicals.<br>• Forecast styles and volumes and contract.<br>• Fair Trade cotton fiber in advance to ensure supply. |
| Promote colored cotton. | • Reduces the overall volume of dyes (and associated pollution), since it can be used undyed. | • Higher price, less availability. Delicate fiber. Limited supply chain participants (growers, ginners, spinners). Forecasting required to ensure supply. Limited marketing opportunity due to low consumer awareness and limited colors. |
| Promote the use of GOTS-certified cotton. | • Organic farming uses crop rotation and biological means to control pests and pathogens. Builds organic matter in the soil; reduces soil, air, and water contamination; increases soil fertility and biodiversity; reduces human and wildlife health hazards; and reduces greenhouse gas emissions resulting from the production of synthetic fertilizers and pesticides.[29] In some regions, organic cotton farmers may gain more income, though this is by no means universal.<br>• Strong marketing potential as there is growing consumer awareness of GOTS certification. | • Higher price. Attaining a GOTS certification takes time and investment. |
| Promote the use of transitional organic cotton. | • Transitional organic cotton uses non-GM seed, is grown in the same manner as organic but does not yet have an organic certification.<br>• Transitional cotton creates a bridge for farmers to switch from conventional to organic cultivation. May be cheaper than organic. | • Cannot claim "organic." Cannot claim transitional organic on marketing material in the United States. |
| Experiment with blends of different types of cottons. | • Balances out cost, aesthetics, and availability. | • Might not be able to make marketing claims, such as "organic" if blends are being used. |

**Figure 1.4**
Sweater made of 80 percent recycled cotton, 20 percent virgin cotton by Vaute, vautecouture.com/.
*Photo by Balarama Heller for VAUTE*

**BOX 1.5**
**FASHION APPLICATIONS—RECYCLED COTTON**

Post-industrial waste is reused by the textile industry and spun into yarns with new apparel applications. Because of its short fiber length, recycled cotton is normally blended with virgin cotton or synthetic fiber to help facilitate processing and add strength to the yarn.

Naturally colored cotton is available in Turkey, Peru, China, Romania, and the United States.

Recycled cotton is available globally, at suppliers such as Pratibha Syntex in India. Availability of recycled cotton for textiles is limited today; availability is likely to increase as virgin resources and landfills become more restricted and new recycling innovations are developed.

Several projects and sources for emerging cotton recycling technologies are in production in the United States, Sweden, Japan, and Europe.

♦ Visit **Part 4: Recycled/Circular Textiles** for more information on these emerging technologies.

# Fashion Applications

- Approximately 60 percent of cotton fiber is used in clothing, most notably in shirts, T-shirts, jeans, coats, jackets, underwear, and foundation garments.[30]
- Organic and transitional cotton is the same quality as conventional cotton and appropriate for any fabric.
- Popular organic cotton products include baby's and children's wear, women's wear, intimate wear/underwear.
- Designing fashion products that command a higher premium is necessary for organic and Fair Trade cotton products.
- While colored cotton can be grown in a range of colors, it is currently being produced only in a narrow range of brown and green colors but can be grown in a variety of other colors at the request of a farmer–partner.
- **Post-industrial waste** is reused by the textile industry and spun into yarns with new apparel applications. Because of its short fiber length, recycled cotton is normally blended with virgin cotton or synthetic fiber to help facilitate processing and add strength to the yarn.
- Post-consumer waste is mostly reused in nonwovens and felts, insulation materials, linings, furniture padding and filling, and paper manufacturing.

# Potential Marketing Opportunities

- **Made from 100 percent organic cotton:** This is acceptable if this is verified accurate.
- **Made with organically grown cotton:** For products that contain 95 percent or more organic cotton, as long as the remaining content is not cotton.
- **Biodegradable:** All fibers, yarns, trims, and dyes used to manufacture the product or garment must also be biodegradable or disassembled before disposal. This should be substantiated with documentation that the product can completely break down into nontoxic material by being processed in a facility where compost is accepted. Secondary label or marketing material should be provided to instruct the customer.
- **Cleaner Cotton™:** If Cleaner Cotton™ is used, consult with the Sustainable Cotton Project to identify the best way to communicate language through marketing and hangtags.
- **Better Cotton:** If Better Cotton is used, consult with the Better Cotton Initiative to identify the best way to communicate language through marketing and hangtags.

- **GOTS Certified Organic:** If certified under GOTS certification, and contains at least 95 percent organic content. Consult with GOTS to identify the best way to communicate language through marketing and hangtags. Do not claim "organic T-shirt" unless all fibers, yarns, trims, and dyes used to manufacture the garment meet GOTS certification standards.
- **Organic In-Conversion:** Can claim in Europe if verified. Be sure to explain on hangtags and POS the story behind organic in-conversion, including that the farmer is working toward organic certification.
- **Naturally colored cotton:** Communicate that naturally colored cotton has been used, and don't forget to tell customers that water, energy, and potential impacts to the environment have been diverted because no dyes have been used.
- **Fair Trade:** Can be claimed if used and verified.
- **OEKO-TEX® certified:** Can be claimed if used and verified.
- **Post-industrial recycled cotton:** Can be claimed if used and verified. Regulations require the percentage of post-consumer and post-industrial recycled content to be stated clearly.
- **Post-consumer recycled cotton:** Can be claimed if used and verified. Regulations require the percentage of post-consumer and post-industrial recycled content to be stated clearly.

# INNOVATION EXERCISES

## Design

1. Design a garment that emphasizes one or more of cotton's natural sustainability benefits.
2. Design a garment that circumvents one or more of cotton's environmental impacts.
3. Design innovative fabrications that use blends with cotton and other natural fibers.
4. Design a garment using 100 percent organic cotton that would command a higher retail price.
5. Design a collection using both cotton and colored cotton while reducing overall volume of dyes (and associated pollution).
6. Design products that feature the unique, heathered, or tweed characteristics of recycled cotton.
7. Design a garment to encourage handwashing or spot cleaning.
8. Create a cotton product that is 100 percent biodegradable along with its constituent parts: dyes, trims, and thread. Or have a disassembly strategy where non-biodegradable elements are disassembled and reused.
9. Design garments that use stripes with recycled cotton blends and virgin cotton for heather/solid effects.
10. Develop a collection of garments that are designed to be easily deconstructed to enable a retail take-back and recycling program. Experiment with seaming and a variety of disassembly mechanisms in different fabrics.

## Merchandising

11. Create a fifty-word marketing communication message for consumers about the advantages of organic cotton or one of the certifications associated with cotton.

## Innovation Exercise Example

**Design a garment that features the unique, heathered, or tweed characteristics of recycled cotton.**

**Figure 1.5**
The design in this example shows a 40 percent recycled cotton/60 percent virgin cotton knit top and bottom.
*Amy Williams*

12. Develop a table that a retail buyer could use to quickly compare the advantages and disadvantages of 100 percent conventional cotton, 100 percent organic cotton, and 100 percent colored cotton.
13. Create a retail point-of-purchase sign or hangtag to promote fashion merchandise made from 100 percent organic cotton and/or naturally colored cotton.
14. Go online and find three fashion retailers who sell fashion merchandise made from 100 percent organic cotton and/or naturally colored cotton. What are the retailers and fashion merchandise offered? Why did they select this fiber for their fashion merchandise? How do they promote the advantages of this fiber to consumers? Do you believe their promotional strategies are accurate and effective? Why or why not?
15. Develop a marketing research protocol (e.g., focus group protocol, online survey, observational strategy) to determine consumers' perceptions of organic cotton and/or colored cotton for fashion merchandise.
16. Develop a retail take-back and recycling program for cotton garments—including the objectives of the program and implementation strategies.

# REVIEW AND RECOMMENDATION

### Not Preferred

Cotton that is conventionally grown with the use of pesticides, herbicides, and fertilizers.

### Preferred

Cotton that has been grown without the use of harmful pesticides is preferred.

- Organic cotton
- Fair Trade cotton
- Integrated Pest Managed cotton

### Recommended Improvements

- Incorporate environmental dyeing practices
- Develop end-of-use strategy (compostability or recyclability)

# OTHER SOURCES


1. Caldwell, Dave. Perspectives Online. "When Roundup Ready Cotton Isn't Ready for Roundup." Spring 2002. Accessed March 16, 2015. www.cals.ncsu.edu/agcomm/magazine/spring02/whenroun.htm
2. CUESA. "From Farm to Garment: Growing a Sustainable Fibershed." Accessed March 16, 2015. www.cuesa.org/article/farm-garment-growing-sustainable-fibershed
3. Food and Agriculture Org. "Natural Fibers Cotton." 2009. Accessed March 16, 2015. www.naturalfibres2009.org/en/fibres/cotton.html

Derek Hall/Getty Images

# Flax

Flax is a natural bast fiber (fiber collected from bast or skin of the plant), along with hemp, ramie, jute, and bamboo. Flax fiber has a high natural luster, and its natural color ranges from beige to light tan to gray. It is a fast and easy growing annual, which requires a cool and relatively humid climate. Linen is the fabric derived from the flax plant.

# Sustainable Benefits

- Flax is a good rotation crop, grows quickly, and requires few chemical inputs in its cultivation. See Part 1 Table 1 for a chart of fast-growing bast fibers.
- Some sources state that flax production requires half the volume of pesticides per acre compared to conventional cotton.[1]
- Flax is a rain-fed crop and generally doesn't require **irrigation**, where water has to be supplied to the land by means of ditches, pipes, or streams.[2]
- Flax may be grown organically without the use of toxic pesticides, and when claimed "**organic**" must meet the standard certification requirements by an internationally recognized certification agency accredited by **International Federation of Agriculture Movements (IFOAM)**.
- Flax fiber has a high natural luster, and its natural color ranges from beige to light tan to gray.
- Once the fiber is extracted from the stem, processing flax into yarn is largely mechanical, with minimal environmental impact, since no chemicals are required.
- Flax fiber and the resulting linen fabric have unique thermo-regulating properties, providing insulation in the winter and good breathability and a cool feeling in the summer.[3]
- In 100 percent form, linen from flax fabric is biodegradable and can be broken down into simpler substances by microorganisms, light, air, or water in a nontoxic process. Absolute biodegradability depends on the dyes and trims used and route of disposal.

    ♦ Visit **Appendix: Biodegradability in the Fashion Fibers STUDIO** for information about designing for biodegradability.

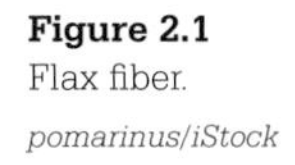
**Figure 2.1**
Flax fiber.
*pomarinus/iStock*

**BOX 2.1 SUSTAINABLE BENEFIT**

Flax fiber and the resulting linen fabric have unique thermo-regulating properties.

# Potential Impacts

## Chemical Use in Cultivation

Flax does require herbicides to control weeds and, as a cellulosic fiber, it also requires some fertilizers. Synthetic fertilizers contain nitrogen salts that salinate the soil and over the long term decrease the productivity of the soil and may pollute aquatic ecosystems.

## Water and Chemicals Used for Extracting Fiber from Plant

Flax is a bast fiber and is extracted directly from the stalk of the plant in a process similar to that used for jute, hemp, and linen made from bamboo. The fiber is extracted through a process called **retting**, which separates the fiber from the stems using microorganisms and moisture. This is carried out in the field (as with **dew retting**) or in tanks (water or chemical retting). Dew retting is preferred, as it utilizes the natural moisture of dew, but it is the longest process, taking over two to three weeks to break down the stems. Although chemical retting is the fastest process, the wastewater is concentrated and rich in chemicals and biological matter and must be treated before its release; otherwise it can harm aquatic ecosystems and receiving water bodies and drinking water sources.[4] See Part 1 Table 2 for more detailed information about the advantages and impacts of different types of retting.

**Figure 2.2**
Dew retting field of flax plants.
*isogood/iStock*

**BOX 2.2 RETTING**

Retting is the process of extracting the fiber from the stalk of the plant, which happens through dew retting, water retting, or chemical retting. Dew retting uses the natural moisture of dew, but is the longest process, taking over two to three weeks to break down the stems.

### Dyeing

The natural color of flax fiber is beige, and flax yarn or fabric must be bleached with chlorine to render it light enough to receive dyes for light or clear shades. Chlorine bleach can form halogenated organic compounds in the wastewater. These compounds bioaccumulate in the food chain, are known **teratogens** (can cause birth defects) and **mutagens** (can change genetic material), are suspected **human carcinogens** (cancer causing), and cause reproductive harm.

- Visit **Chapter 22: Bleaching** for information on the environmental impacts of bleaching.
- Visit **Chapter 23: Dyeing and Printing** for more information on impacts associated with dyeing.
- Visit **Chapter 26: Recycled/Circular Textiles Technologies** for more information about recyclability.
- Visit **Appendix B: Consumer Care and Washing** for more information.
- Visit **Appendix: Biodegradability in the Fashion Fibers STUDIO** for information about designing for biodegradability.

### Consumer Care/Washing

Flax may be machine washed or dry-cleaned. Since the fabric wrinkles easily, it often requires heavy pressing to render it smooth after wash. To iron a shirt for fifteen minutes once every other week would require 7.15 kWh annual energy consumption, which at 11 cents per kWh is about 79 cents per year.[5] While it is important to know that a flax garment will require heavy pressing and electricity use in its lifetime, the amount of electricity is relatively low, and it might be useful to instead address other potential impacts that could have more influence.

- Visit **Appendix B: Consumer Care and Washing** for more information.

## Availability

Seventy percent of the world's crop is produced in Europe, and 10,000 companies from fourteen European countries cover all stages of the fiber's production and transformation.[6]

Dew-retted flax linen is readily available, although it has to be specially requested. Certification on the retting method used should also be requested.

Organic flax linen is less available and more expensive than conventional flax linen. Organic certification by an internationally recognized certification agency accredited by IFOAM must be in place.

**Table 2.1 Optimize the Sustainability Benefits of Flax**

| Opportunity | Benefit | Considerations |
|---|---|---|
| Promote suppliers using organic flax. | • Ensures that no disallowed fertilizers are used. | • Organic certification must be in place by a recognized international certification agency accredited by **IFOAM**.<br>• Organic flax linen is not as readily available as conventional flax linen and commands a premium. |
| Promote the use of natural color. | • No bleaches or dyes are used in this case, and associated pollution impacts are avoided. | • Using the natural color of flax linen will limit your color palette to neutrals. Spice up a neutral color palette with interesting drape, silhouettes, or seaming. |
| Promote the use of non-chlorine bleaches, such as hydrogen peroxide, to lighten the natural beige color for dyeing dark shades and bright/light shades.<br>♦ Visit **Chapter 22: Bleaching** for more guidance. | • Hydrogen peroxide harmlessly **decomposes** into water and oxygen gas. | • Non-chlorine bleaches do not strip out the original color of the fiber. Consequently, lighter and brighter colors will be duller due to the overdyed effect. Non-chlorine bleaching is adequate for dark colors, which mask the original beige tone. |
| Promote the use of **ozone** bleaching processes, which use ozone gas (a much more powerful oxidizing agent than chlorine) to strip out the natural beige color of flax linen.<br>Promote the particular aesthetic of ozone bleach effects.<br>♦ Visit **Chapter 22: Bleaching** for more guidance. | • It is very effective at fading pre-dyed/preprinted fabrics and garments, can be performed at lower temperatures than with hydrogen peroxide, and uses no water at all. In addition, ozone is completely biodegradable, as it reverts rapidly back to oxygen ($O_2$), leaving no chemical residue. | • Ozone has limited availability and is relatively expensive since it requires investment in ozone-generating equipment.<br>• Ozone processes produce a different aesthetic than chlorine derivative or permanganate bleaching. |
| Promote the use of **enzymes** to strip out the natural beige color of flax linen.<br>Promote the particular aesthetic of enzyme bleaches. | • Enzymes use less water than bleaching.<br>• Enzymes are fully biodegradable.<br>• Enzymes are gentler to the fiber than bleaching and do not cause degradation as bleaching can.<br>• The wastewater for enzymes is easier and cheaper to clean than water from bleaching. | • Enzymes are not allowed under **Global Organic Textile Standard (GOTS)** standards because most technical enzymes are produced with genetically modified organisms (GMOs). GMOs and enzymes are not allowed for use in products labeled as "organic" or "made with organic," according to GOTS standard.<br>• Enzymes produce a different aesthetic than chlorine derivative or permanganate bleaching. |
| Promote suppliers who use dew retting over water or chemical retting. | • Dew retting reduces the biological load in the receiving water bodies and adds nutrients to the soil. | • The natural color may vary slightly from lot to lot, since the process is influenced by weather.<br>• Dew retting takes longer than the other types of retting—usually over two to three weeks to break down the stems. |
| Promote suppliers who use enzymatic retting over water or chemical retting. | • The process is faster and leaves the water unharmed. Enzymatic retting can be commercially reproduced. | • Low fiber strength. The process is less common than other retting processes. |
| Actively seek out stain-resistant finishes for flax linen. | • Will reduce washing, ironing, and dry cleaning by the consumer, and the water and pollution associated with consumer care. | • Ensure that stain-resistant finishes are nontoxic to air and water.<br>♦ Visit **Chapter 24: Finishing** for more information. |

**BOX 2.3**
**FASHION APPLICATIONS**

In 100 percent fabrications, flax linen is durable and especially ideal for jackets and pants.

**Figure 2.3**
Flax linen men's suit.
*Catwalking/Getty Images*

# Fashion Applications

In 100 percent fabrications, flax linen is durable and available in a variety of yarn counts and fabric types. Suitable product applications include jeans, dress pants, jackets, dress shirts, handkerchief-weight blouses, knits, bed linens, and outdoor fabrics. Blends of cotton/flax linen are machine washable and suitable for sportswear, wovens, and knits.

# Marketing Opportunities

- **Fast-growing natural resource:** Flax is considered a fast-growing resource, as seen in Part 1 Table 1, and can be marketed as such.
- **Low water footprint in cultivation:** Flax is biologically efficient, mainly rain fed, and can grow in diverse climates.
- **Organic:** If organic flax fiber is used. All fibers, yarns, trims, and dyes used to manufacture the garment must comply with the GOTS organic garment standard. Simply state "made from 100 percent organic flax," if this is verified and accurate.

- **Biodegradable:** All fibers, yarns, trims, and dyes used to manufacture the product or garment must also be biodegradable, or disassembled before disposal. This should be substantiated with documentation that the product can completely break down into nontoxic material by being processed in a facility where compost is accepted. A secondary label or marketing material should be provided to instruct customer.
- **Non-chlorine bleached:** If alternative bleach is used. Specify alternative bleaching method and explain environmental benefits on secondary label or marketing material.
- **Dew retted:** If dew retted processed.
- **Enzyme retted:** If enzyme retted processed.
- **Renewable resource:** Flax is a renewable resource and can be replaced or replenished by natural processes.

# INNOVATION EXERCISES

## Design

1. Design a garment that emphasizes one or more of flax linen's sustainability benefits.
2. Design a garment that circumvents one or more of flax linen's environmental impacts.
3. Design a garment that uses flax linen fiber in blends with cotton to negate the need for using chlorine bleach.
4. Design a season-less garment that allows for year-round wear, leveraging flax fibers' unique thermo-regulating properties—providing insulation in the winter and good breathability and a cool feeling in the summer.
5. Design a "culturally durable" (i.e., styling that doesn't date with passing trends) garment in flax linen to optimize the fiber's physical durability.
6. Design a garment that utilizes the natural wrinkling of the fabric as a design feature to influence the customer to reduce ironing and energy use.
7. Create a flax product that is 100 percent biodegradable: the product can break down in a reasonable amount of time and is equipped with an ingredient that provides valuable nutrients to the soil after disposal.
8. Design a garment to encourage handwashing or spot cleaning.

## Merchandising

9. Create a fifty-word marketing communication message for consumers about the sustainability benefits of flax.
10. Incorporating information from chapters 1 and 2, develop a table that a retail buyer could use to quickly compare the advantages and disadvantages of 100 percent conventional cotton, 100 percent organic cotton, and 100 percent flax.
11. Create a retail point-of-purchase sign or hangtag to promote fashion merchandise made from 100 percent flax.
12. Go online and find three fashion retailers who sell fashion merchandise made from 100 percent flax. What are the retailers and fashion merchandise offered? Why did they select this fiber for their fashion merchandise? How do they promote the advantages of this fiber to consumers? Do you believe their promotional strategies are accurate and effective? Why or why not?

## Innovation Exercise Example

**Design a season-less garment that allows for year-round wear, leveraging flax fibers' unique thermo-regulating properties—providing insulation in the winter and good breathability and a cool feeling in the summer.**

**Figure 2.4**
The design in this example shows a 100 percent linen from flax medium-weight woven jacket for fall with removable sleeves to create a vest for spring-time wear.
*Amy Williams*

13. Develop a marketing research protocol (e.g., focus group protocol, online survey, observational strategy) to determine consumers' perceptions of flax for fashion merchandise.

# REVIEW AND RECOMMENDATION

## Preferred

Flax for linen fiber is preferred for use as a sustainable fiber because:

- Flax is a fast-growing renewable fiber that is annually farmed.
- Flax is a good rotation crop and is a biologically efficient, low-maintenance crop that requires few chemical inputs during the growing season. It is mainly rain fed, traditionally farmed, and grown similarly to organic produce.[7]
- Flax is a rain-fed crop and generally doesn't require irrigation, where water has to be supplied to the land by means of ditches, pipes, or streams.
- Once the fiber is extracted from the stem, processing flax into yarn is largely mechanical, with minimal environmental impact, since no chemicals are required.
- Flax fiber and the resulting linen fabric have unique thermo-regulating properties, providing insulation in the winter and good breathability and a cool feeling in the summer.
- Flax is biodegradable.

## Recommended Improvements

- Flax fiber that has been either dew retted or enzyme retted, as it reduces the biological load in the receiving water bodies
- Flax fiber that has been bleached with alternative methods to prepare the fiber for dyeing, such as ozone bleaching, hydrogen peroxide, or enzyme processes
- Incorporating environmental dyeing practices
- Developing end-of-use strategy (whether compostability or recyclability)
- Flax that has been grown organically

fotohunter/iStock

# Bamboo Linen

Bamboo is a natural bast fiber (fiber collected from bast or skin of the plant), along with hemp, ramie, jute, and flax. Bamboo can be processed to make two types of fiber: bamboo linen and bamboo rayon/viscose.

Bamboo linen is a natural fiber, and processing into yarn is largely mechanical with minimal environmental impact.

Bamboo rayon/viscose is a manufactured cellulosic fiber, which has to undergo chemically intensive processing to make it soft to the touch.

# Sustainable Benefits

- Bamboo is a "**rapidly renewable**" resource, meaning that it grows quickly and can be harvested at least once a year.[1,2] See Part 1 Table 1 for a chart of fast-growing plant fibers.
- Bamboo linen has a natural ability to breathe and wick moisture away due to its **porous** nature. It keeps the wearer cooler—by one to two degrees—than someone wearing cotton.
- Bamboo is a biologically efficient, low-maintenance crop that requires few chemical inputs during the growing season. It is mainly rain fed and can grow in diverse climates.
- Due to its speedy growth and little input needed for growing, some say that using bamboo as an alternative to wood trees could help slow deforestation.[3]
- Once the fiber is extracted from the stem, processing bamboo into yarn for linen is largely **mechanical**, with minimal environmental impact.
- In 100 percent form, linen from bamboo fiber is **biodegradable** and can be broken down into simpler substances by microorganisms, light, air, or water in a nontoxic process. Absolute biodegradability depends on the dyes and trims used and route of disposal.

  ♦ Visit **Appendix: Biodegradability in the Fashion Fibers STUDIO** for information about designing for biodegradability.

**Figure 3.1**
Bamboo plantation.
*Betty4240/iStock*

**BOX 3.1 SUSTAINABLE BENEFIT**

Due to its speedy growth and little input needed for growing, some say that using bamboo as an alternative to wood trees could help slow deforestation.[4]

# Potential Impacts

## Water and Chemicals Used for Extracting Fiber from Plant

When processing for linen, bamboo is a bast fiber and is extracted directly from the stalk of the plant in a process similar to that used for jute, hemp, and flax. The fiber is extracted through a process called **retting**, which separates the fiber from the stems using microorganisms and moisture. This is carried out in the field (as with dew retting) or in tanks (in water or chemical retting). Dew retting is preferred, as it utilizes the natural moisture of dew, but it is the longest process, taking over two to three weeks to break down the stems. Part 1 Table 2 details different retting processes. Although chemical retting is the fastest process, the wastewater is concentrated and rich in chemicals and biological matter and must be treated before its release; otherwise, it can harm aquatic ecosystems and receiving water bodies and drinking water sources.[5] See Part 1 Table 2 for more detailed information about the advantages and impacts of each type of retting.

## Dyeing

The natural color of bamboo fiber is golden, and bamboo linen must be bleached with chlorine to render it light enough to receive dyes for light or clear shades. Chlorine bleach can form **halogenated organic compounds** in the wastewater. These compounds **bioaccumulate** in the food chain, are known **teratogens** (can cause birth defects) and **mutagens** (can change genetic material), are suspected **human carcinogens** (cancer causing), and cause reproductive harm.

- Visit **Chapter 22: Bleaching** for information on the environmental impacts of bleaching.

**Figure 3.2**
Bamboo fiber.
*Yelena Safranova*

**BOX 3.2 SUSTAINABLE BENEFIT**

The natural color of bamboo fiber is golden, and bamboo linen must be bleached with chlorine to render it light enough to receive dyes for light or clear shades.

- Visit **Chapter 23: Dyeing and Printing** for more information on impacts associated with dyeing.
- Visit **Chapter 26: Recycled/Circular Textiles Technologies** for more information about recyclability.
- Visit **Appendix B: Consumer Care and Washing** for more information.
- Visit **Appendix: Biodegradability in the Fashion Fibers STUDIO** for information about designing for biodegradability.


### Consumer Care/Washing

Bamboo linen may be machine washed or dry-cleaned. Since the fabric wrinkles easily, it often requires heavy pressing to render it smooth after wash. To iron a shirt for fifteen minutes once every other week, would require 7.15 kWh annual energy consumption, which at 11 cents per kWH is about 79 cents per year.[6] While it is important to know that bamboo linen garments will require heavy pressing and electricity use in their lifetime, the amount of electricity is relatively low, and it might be useful to instead address other potential impacts that have more influence.


- Visit **Appendix B: Consumer Care and Washing** for more information.


**Table 3.1 Optimize the Sustainability Benefits of Bamboo**

| Opportunity | Benefit | Considerations |
|---|---|---|
| Know the difference between bamboo linen fabric and bamboo made from a viscose process. | • Ensures that accurate and honest marketing claims are being made and that your design decisions are based on accurate sustainability information. | • Bamboo viscose is **chemically processed** and has greater impacts to water and air.<br>♦ Visit **Chapter 18: Rayon/Viscose (Made from Bamboo)** for more information. |
| Promote the use of linen from bamboo products. | • Once the fiber is extracted from the stem, processing bamboo into yarn for linen is largely mechanical, with minimal environmental impact. | • Not all apparel applications will be suitable for bamboo linen.<br>• Bamboo linen is not as readily available as other fibers, including bamboo rayon/viscose, and may be more difficult to find and more expensive than alternatives. |
| Promote suppliers using organic bamboo. | • Ensures that no disallowed fertilizers are used. | • Organic certification must be in place by a recognized international certification agency accredited by **IFOAM**.<br>• Organic linen from bamboo is not as readily available as conventional linen from flax and commands a premium. |
| Promote the use of natural color. | • No bleaches or dyes are used in this case, and associated pollution impacts are avoided. | • Using the natural color of bamboo linen will limit your color palette to neutrals. Spice up a neutral color palette with interesting drape, silhouettes, or seaming. |
| Promote suppliers who use dew retting over water or chemical retting. | • Dew retting reduces the biological load in the receiving water bodies and adds nutrients to the soil. | • The natural color may vary slightly from lot to lot, since the process is influenced by weather. |

| Opportunity | Benefit | Considerations |
|---|---|---|
| Promote the use of non-chlorine bleaches, such as **hydrogen peroxide**, to lighten the natural beige color for dyeing dark shades and bright/light shades.<br>♦ Visit **Chapter 22: Bleaching** for information on these alternatives. | • Hydrogen peroxide harmlessly decomposes into water and oxygen gas. | • Non-chlorine bleaches do not strip out the original color of the fiber. Consequently, lighter and brighter colors will be duller due to the overdyed effect. Non-chlorine bleaching is adequate for dark colors, which mask the original beige tone. |
| Promote suppliers who use enzymatic retting over water or chemical retting. | • The process is faster and leaves the water unharmed. Enzymatic retting can be commercially reproduced. | • Low fiber strength. The process is less common than other retting processes. |
| Promote the use of **ozone** bleaching processes, which use ozone gas (a much more powerful oxidizing agent than chlorine) to strip out the natural beige color of bamboo.<br>Promote the particular aesthetic of ozone bleach effects.<br>♦ Visit **Chapter 22: Bleaching** for information on these alternatives. | • It is very effective at fading pre-dyed/preprinted fabrics and garments, can be performed at lower temperatures than with hydrogen peroxide, and uses no water at all. In addition, ozone is completely biodegradable, as it reverts rapidly back to oxygen ($O_2$), leaving no chemical residue. | • Ozone has limited availability and is relatively expensive since it requires investment in ozone-generating equipment.<br>• Ozone processes produce a different aesthetic than chlorine derivative or permanganate bleaching. |
| Promote the use of enzymes to strip out the natural beige color of bamboo.<br>Promote the particular aesthetic of enzyme bleaches. | • Enzymes use less water than bleaching.<br>• Enzymes are fully biodegradable.<br>• Enzymes are gentler to the fiber than bleaching and do not cause degradation as bleaching can.<br>• The wastewater for enzymes is easier and cheaper and cleaner than bleaching. | • Enzymes are not allowed under **Global Organic Textile Standard (GOTS)** standards because most technical enzymes are produced with genetically modified organisms (GMOs). GMOs and enzymes are not allowed for use in products labeled as "organic" or "made with organic," according to GOTS standard.<br>• Enzymes produce a different aesthetic than chlorine derivative or permanganate bleaching. |
| Promote the use of bamboo linen–blended fabrics. | • Can achieve the property benefits of both fibers. | • This sometimes requires further processing, which can include chemicals. When blended with a synthetic fiber, bamboo linen will no longer be biodegradable. |

# Availability

Bamboo linen fabric is available from Deltracon in Belgium. This company produces a heavier quality bamboo linen fabric suitable for home furnishings.

*Expressing an interest in bamboo linen fabric to your suppliers can help to expand its availability worldwide.*

# Fashion Applications

Linen from bamboo can be used in whatever application best suits the styling and function of the garment designed. Linen from bamboo fabrics is seen in both knits and wovens, and range from medium-weight jerseys to heavyweight wovens for trousers.

**BOX 3.3**
**FASHION APPLICATIONS**

While linen from bamboo is seen in both knits and wovens, it is ideal for button-up shirt applications, as it will drape beautifully and maintain a crispness (although it will wrinkle easily!).

**Figure 3.3**
Bamboo linen shirt.
*Chris Jackson/Getty Images*

# Potential Marketing Opportunities

- **Bamboo linen:** The Federal Trade Commission (FTC) in the United States provides a conservative approach to labeling bamboo products and may be used as a guide. The FTC requires that fabric should not be referred to simply as "bamboo," since the processing of bamboo into linen and into rayon/viscose varies significantly. The bamboo textile should be referred to more specifically as "bamboo linen" or "rayon made from bamboo." This should be done consistently on labeling, hangtags, and point-of-sale (POS) materials.[7]
- **Fast-growing natural resource:** Bamboo is considered a fast-growing resource, as seen in Part 1 Table 1, and can be marketed as such.
- **Low water footprint in cultivation:** Bamboo is biologically efficient, is mainly rain fed, and can grow in diverse climates.
- **Organic:** If organic bamboo fiber is used, then all fibers, yarns, trims, and dyes used to manufacture the garment must comply with GOTS. Label as "made from 100 percent organic bamboo" if this is verified and accurate.
- **Biodegradable:** Undyed bamboo fiber is biodegradable. If claiming a garment is biodegradable, all fibers, yarns, trims, and dyes used to

manufacture the product or garment must also be biodegradable or disassembled before disposal. This should be substantiated with documentation that the product can completely break down into nontoxic material by being processed in a facility where compost is accepted. Secondary labeling or marketing material should be provided to instruct the customer.

- **Non-chlorine bleached:** If alternative bleach is used. Specify the alternative bleaching method and explain the environmental benefits on the secondary label or marketing material.
- **Dew retted:** If dew retted processed.
- **Enzyme retted:** If enzyme retted processed.
- **Antibacterial:**[8] The antibacterial properties may depend on the manufacturing process, with only mechanically processed bamboo (bamboo linen) providing a benefit.[9]
- **Renewable resource:** Bamboo is a renewable resource and can be replaced or replenished by natural processes.

# INNOVATION EXERCISES

## Design

1. Design a garment that emphasizes one or more of bamboo linen's sustainability benefits.
2. Design a garment that circumvents one or more of bamboo linen's environmental impacts.
3. Design a collection that features bamboo linen in 100 percent natural color blended with a natural fiber of your choosing.
4. Design a garment that utilizes the natural wrinkling of the fabric as a design feature to influence the customer to reduce ironing and energy use.
5. Design a completely biodegradable bamboo linen collection—without trims.
6. Design a garment to encourage handwashing or spot cleaning.

## Merchandising

7. Create a fifty-word marketing communication message for consumers about the sustainability benefits of bamboo linen.
8. Incorporating information from chapters 1, 2, and 3 develop a table that a retail buyer could use to quickly compare the advantages and disadvantages of 100 percent conventional cotton, 100 percent organic cotton, 100 percent flax, and 100 percent bamboo linen.
9. Create a retail point-of-purchase sign or hangtag to promote fashion merchandise made from 100 percent bamboo linen.
10. Go online and find three fashion retailers who sell fashion merchandise made from 100 percent bamboo linen. What are the retailers and fashion merchandise offered? Why did they select this fiber for their fashion merchandise? How do they promote the advantages of this fiber to consumers? Do you believe their promotional strategies are accurate and effective? Why or why not?
11. Develop a marketing research protocol (e.g., focus group protocol, online survey, observational strategy) to determine consumers' perceptions of bamboo linen for fashion merchandise.

## Innovation Exercise Example

**Design a garment that utilizes the natural wrinkling of the fabric as a design feature to influence the customer to reduce ironing and energy use.**

**Figure 3.4**
One hundred percent linen-from-bamboo dress with gathering below the waist to reduce visual signs of wrinkling from sitting and movement.
*Amy Williams*

# REVIEW AND RECOMMENDATION

## Preferred

Bamboo for linen fiber is preferred for use as a sustainable fiber because:

- Bamboo is a rapidly renewable fiber; it can grow quickly and can be harvested at least once a year.
- Bamboo linen has a natural ability to breathe and wick moisture away due to its porous nature. It keeps the wearer cooler—by one to two degrees—than someone wearing cotton.
- Due to its speedy growth and little input needed for growing, some say that using bamboo as an alternative to wood trees could help slow deforestation.[1]
- Bamboo is a biologically efficient, low-maintenance crop that requires few chemical inputs during the growing season. It is mainly rain fed and can grow in diverse climates.
- Once the fiber is extracted from the stem, processing bamboo into yarn for linen is largely mechanical, with minimal environmental impact.
- Bamboo is biodegradable.

## Recommended Improvements

- Bamboo fiber that has been either dew retted or enzyme retted, as it reduces the biological load in the receiving water bodies
- Bamboo that has been grown organically
- Bamboo fiber that has been bleached with alternative methods to prepare the fiber for dyeing, such as ozone bleaching or hydrogen peroxide or enzyme processes
- Incorporating environmental dyeing practices
- Developing end-of-use strategy (whether compostability or recyclability)

Achim Prill/iStock

4

# Hemp

Hemp is a tall, sturdy plant that is grown to make a variety of foods, oils, and textiles. Hemp is a natural bast fiber (fiber collected from bast or skin of the plant), along with jute, ramie, bamboo, and flax.

# Sustainable Benefits

- Hemp is a **rapidly renewable** fiber; it can grow up to 4 meters high in only three months. See Part 1 Table 1 for a growth comparison chart that shows hemp and other fast-growing fibers.
- Hemp is a hardy plant and can be grown in cool climates on marginal land with minimal water. Hemp has the highest yield per acre of any natural fiber.
- Hemp is a low-maintenance crop, which requires few chemical inputs during the growing season. It shows great adaptability to various climactic conditions and requires no herbicides or irrigation water. Once established, hemp is naturally resistant to pests and diseases and therefore does not require pesticide applications. To ensure high nutrient content and guarantee the tall growth of the plant for increased fiber content, it may require the use of fertilizers.[1]
- Hemp may be grown organically, but must meet the certification requirements of an internationally recognized certification agency accredited by the **International Federation of Agriculture Movements (IFOAM)**.
- Its extensive root system allows hemp to use deep supplies of nutrients and water.[2] Since the leaves and outer stalks of the plant are left in the field after harvest, the nitrogen they contain absorbs into the soil and food crops can be grown immediately without having to leave the fields fallow. Though hemp can be grown successfully on the same land for several years, rotation with other crops is desirable.[3]
- Once the hemp fiber is extracted from the stem, processing it into yarn is largely **mechanical** with minimal environmental impact.
- When used by itself, hemp is one of the strongest natural fibers and has always been valued for its durability. It offers good breathability in the summer as well as thermal protection in the winter. Hemp fiber conducts heat, resists mildew, blocks ultraviolet light, and has natural antibacterial properties.[4]

**Figure 4.1**
Undyed hemp fiber.
*Simone Andress /Shutterstock*

# Potential Impacts

### Water and Chemicals Used for Extracting Fiber from Plant

Hemp is a bast fiber and is extracted directly from the stalk of the plant in a process similar to that used for flax and linen (from bamboo). The fiber is extracted through a process called retting, which separates the fiber from the stems using microorganisms and moisture.[5] This is carried out in the field (as with dew retting) or in tanks (in water or chemical retting). Dew retting is preferred, as it utilizes the natural moisture of dew over two to three weeks to break down the stems, and in the process it returns nutrients into the soil.[6] Tank retting is a faster process, taking four to six days. Tank retting allows greater control and produces more uniform quality, but the wastewater from tank retting is concentrated and rich in chemicals and biological matter, which negatively impacts receiving water bodies and harms aquatic ecosystems, if left untreated before its release.[7] Part 1 Table 2 details different retting processes. Although chemical retting is the fastest process, the wastewater is concentrated and rich in chemicals and biological matter and must be treated before its release; otherwise it can harm aquatic ecosystems and receiving water bodies and drinking water sources. See Part 1 Table 2 for more detailed information about the advantages and impacts of each type of retting.

### Dyeing

The natural color of hemp fiber is beige, and hemp yarn or fabric must be bleached with chlorine to render it light enough to receive dyes for light or clear shades. Chlorine bleach can form halogenated organic compounds in the wastewater. These compounds bioaccumulate in the food chain, are known **teratogens** (can cause birth defects) and **mutagens** (can change genetic material), are suspected **human carcinogens** (cancer causing), and cause reproductive harm.

- Visit **Chapter 23: Dyeing and Printing** for more information on impacts associated with dyeing.

### Consumer Care/Washing

Hemp fiber may be machine washed or dry-cleaned. Since the fabric wrinkles easily, it often requires heavy pressing to render it smooth after wash. To iron a shirt for fifteen minutes once every other week, would require 7.15 kWh annual energy consumption, which at 11 cents per kWh is about 79 cents per year.[8] While it is important to know that hemp garments will require heavy pressing and electricity use in their lifetime, the amount of electricity is relatively low, and it might be useful to instead address other potential impacts that have more influence.

- Visit **Chapter 26: Recycled/Circular Textiles Technologies** for more information about recyclability.
- Visit **Appendix B: Consumer Care and Washing** for more information.
- Visit **Appendix: Biodegradability in the Fashion Fibers STUDIO** for information about designing for biodegradability.

**Table 4.1 Optimize the Sustainability Benefits of Hemp**

| Opportunity | Benefit | Considerations |
|---|---|---|
| Promote suppliers using organic hemp. | • In addition to the general ecological benefits of hemp, organic processes ensure that no disallowed fertilizers are used. | • Organic certification must be in place by a recognized international certification agency accredited by **IFOAM**.<br>• Organic hemp is not as readily available as conventional hemp and commands a premium.<br>• Although organic certification disallows the use of chemicals in the growing of hemp, it does not necessarily guarantee low water use, fair labor practices, or a fair price to the farmer. |
| Promote the use of natural color. | • No bleaches or dyes are used in this case, and associated pollution impacts are avoided. | • Using the natural color of hemp will limit your color palette to neutrals. Spice up a neutral color palette with interesting drape, silhouettes, or seaming. |
| Promote suppliers who use dew retting over water or tank retting. | • Dew retting reduces the biological load in the receiving water bodies and adds nutrients to the soil. | • The natural color may vary slightly from lot to lot, since the process is influenced by weather. |
| Use hydrogen peroxide to lighten the natural beige color for dyeing dark shades and bright/light shades.<br>♦ Visit **Chapter 22: Bleaching** for more guidance. | • Hydrogen peroxide harmlessly decomposes into water and oxygen gas. | • Non-chlorine bleaches do not strip out the original color of the fiber. Consequently, colors will be duller due to the overdyed effect. Non-chlorine bleaching is adequate for dark colors, which mask the original beige tone. |
| Promote using ozone bleaching processes, which use ozone gas (a much more powerful oxidizing agent than chlorine) to strip out the natural beige color of hemp.<br>Promote the particular aesthetic of ozone bleach effects. | • It is very effective at fading pre-dyed/pre-printed fabrics and garments, can be performed at lower temperatures than with hydrogen peroxide, and uses no water at all. In addition, ozone is completely biodegradable, as it reverts rapidly back to oxygen ($O_2$), leaving no chemical residue. | • Ozone has limited availability and is relatively expensive since it requires investment in ozone-generating equipment.<br>• Ozone processes produce a different aesthetic than chlorine derivative or permanganate bleaching. |
| Promote the use of enzymes to strip out the natural beige color of hemp.<br>Promote the particular aesthetic of enzyme bleaches. | • Enzymes use less water than bleaching.<br>• Enzymes are fully biodegradable.<br>• Enzymes are gentler to the fiber than bleaching and do not cause degradation as bleaching can.<br>• The wastewater for enzymes is easier and cheaper to treat than bleaching. | • Enzymes are not allowed under **Global Organic Textile Standard (GOTS)** standards because most technical enzymes are produced with genetically modified organisms (GMOs). GMOs and enzymes are not allowed for use in products labeled as "organic" or "made with organic," according to GOTS standard.<br>• Enzymes produce a different aesthetic than chlorine derivative or permanganate bleaching. |

# Availability

Hemp is readily available in 100 percent as well as blends with cotton and silk.

A number of farmers in the United Kingdom, Canada, and Denmark are currently growing organic hemp.[9] Organic certification by an internationally recognized certification agency accredited by IFOAM must be in place.

# Fashion Applications

As 100 percent hemp is highly durable and suitable for workwear items and sportswear from jeans to jackets.

Fine-quality hemp fabrics can be comparable to linen, with a similar look and feel. Combining hemp with other fibers improves the hand of the fabric.[10]

**Figure 4.2**
Fifty-five percent hemp, 45 percent organic cotton by Nurmi Clothing, www.nurmiclothing.com/.
*Antti Ahtiluoto*

**BOX 4.1**
**FASHION APPLICATIONS**

As 100 percent, hemp is highly durable and suitable for workwear items and sportswear from jeans to jackets.

In fabric blends, hemp fiber is cut into shorter pieces to be able to process on a cotton system. Depending on the percentage of hemp in the blend, the fabric assumes the characteristics of the other fiber. Silk hemp blends are drapey and suitable for dress shirts, dresses, and even underwear. Cotton hemp blends are washable and suitable for sportswear, wovens, and knits.

# Potential Marketing Opportunities

- **Fast-growing natural resource:** Hemp is considered a fast-growing resource, as seen in Part 1 Table 1, and can be marketed as such.
- **Low water footprint in cultivation:** Hemp is biologically efficient, mainly rain fed, and can grow in diverse climates.
- **Organic:** If organic hemp fiber is used. All fibers, yarns, trims, and dyes used to manufacture the garment must comply with the GOTS. Simply state "made from 100 percent organic hemp," if this is verified and accurate.
- **Biodegradable:** All fibers, yarns, trims, and dyes used to manufacture the product or garment must also be biodegradable or disassembled before disposal. This should be substantiated with documentation that the product can completely break down into nontoxic material by being processed in a facility where compost is accepted. Secondary labeling or marketing material should be provided to instruct the customer.
- **Non-chlorine bleached:** If alternative bleach is used. Specify the alternative bleaching method and explain the environmental benefits on the secondary label or marketing material.
- **Dew retted:** If dew retted processed.
- **Enzyme retted:** If enzyme retted processed.
- **Renewable resource:** Hemp is a renewable resource and can be replaced or replenished by natural processes.

# INNOVATION EXERCISES

## Design

1. Design a garment that emphasizes one or more of hemp's sustainability benefits.
2. Design a garment that circumvents one or more of hemp's environmental impacts.
3. Design a "**culturally durable**" (i.e., styling that doesn't date with passing trends) garment in hemp to optimize the fiber's physical durability.
4. Design a garment where hemp is strategically placed in areas of high stress to maximize its physically durable properties.
5. Design a garment that would traditionally be made with cotton. After all, the original jean was made out of hemp!
6. Design a garment that utilizes the natural wrinkling of the fabric as a design feature to influence the customer to reduce ironing and energy use.

## Innovation Exercise Example

**Design a garment that would traditionally be made with cotton.**

**Figure 4.3**
The design in this example shows a 100 percent cotton hemp jeans. The original jean was made out of hemp!
*Amy Williams*

7. When designing hemp products, it's best to make them classic enough that they will transcend fashion trends, since physical durability is a benefit for the wearer but a liability in the environment when the garment is discarded.

### Merchandising

8. Create a fifty-word marketing communication message for consumers about the sustainability benefits of hemp.
9. Incorporating information from chapters 1, 2, 3, and 4, develop a table that a retail buyer could use to quickly compare the advantages and disadvantages of 100 percent conventional cotton, 100 percent organic cotton, 100 percent flax, 100 percent bamboo linen, and 100 percent hemp.
10. Create a retail point-of-purchase sign or hangtag to promote fashion merchandise made from 100 percent hemp.
11. Go online and find three fashion retailers who sell fashion merchandise made from 100 percent hemp. What are the retailers and fashion merchandise offered? Why did they select this fiber for their fashion merchandise? How do they promote the advantages of this fiber to consumers? Do you believe their promotional strategies are accurate and effective? Why or why not?
12. Develop a marketing research protocol (e.g., focus group protocol, online survey, observational strategy) to determine consumers' perceptions of hemp for fashion merchandise.

## REVIEW AND RECOMMENDATION

### Preferred

Hemp fiber is preferred for use as a sustainable fiber because:

- Hemp is a rapidly renewable fiber; it can grow up to 4 meters high in only three months.
- Hemp is a biologically efficient, low-maintenance crop that requires few chemical inputs during the growing season. It is mainly rain fed, traditionally farmed, and grown similarly to organic produce.[11]
- Hemp is a hardy plant and can be grown in cool climates on marginal land with minimal water. Hemp has the highest yield per acre of any natural fiber.
- Hemp shows great adaptability to various climactic conditions and requires no herbicides or irrigation water. Once established, hemp is naturally resistant to pests and diseases and therefore does not require pesticide applications.
- Its extensive root system allows hemp to use deep supplies of nutrients and water.[12] Since the leaves and outer stalks of the plant are left in the field after harvest, the nitrogen they contain absorbs into the soil and food crops can be grown immediately without having to leave the fields fallow.
- Once the hemp fiber is extracted from the stem, processing it into yarn is largely mechanical with minimal environmental impact.
- When used by itself, hemp is one of the strongest natural fibers and has always been valued for its durability.

## Recommended Improvements

- Hemp fiber that has been either dew retted or enzyme retted, as it reduces the biological load in the receiving water bodies
- Hemp fiber that has been bleached with alternative methods to prepare the fiber for dyeing, such as ozone bleaching or hydrogen peroxide or enzyme processes
- Incorporating environmental dyeing practices
- Developing end-of-use strategy (whether compostability or recyclability)

Richard I'Anson/Lonely Planet Images/Getty Images

# 5

# Jute

Jute is a long, soft, shiny plant fiber that can be spun into course, strong threads. Jute is one of the most inexpensive natural fibers and is second only to cotton in amount produced and variety of uses.[1] Jute is a natural bast fiber (fiber collected from bast or skin of the plant) along with hemp, ramie, bamboo, and flax.

# Sustainable Benefits

- Jute is a fast-growing renewable fiber that is annually farmed. See Part 1 Table 1 for a growth comparison chart that shows jute and other fast-growing fibers.
- Jute is a **biologically efficient**, low-maintenance crop that requires few chemical inputs during the growing season. It is mainly rain fed, traditionally farmed, and grown similarly to organic produce.[2]
- Jute has a natural luster and is valued for its durability, fair abrasion resistance, and high tensile strength. Jute fiber has antistatic properties,

**Figure 5.1**
Golden jute fiber.
*By shibu bhattacharje/ Moment/Getty Images*

**BOX 5.1 SUSTAINABLE BENEFIT**

Jute has a natural luster and is valued for its durability, fair abrasion resistance, and high tensile strength.

heat insulation, and low elongation, which helps to retain its shape. Jute fiber is colorfast and lightfast.

- Jute may be grown organically, but it must meet the certification requirements of an internationally recognized certification agency accredited by **International Federation of Agriculture Movements (IFOAM).**
- Studies reveal that the $CO_2$ assimilation rate of jute is several times higher than that of trees. During the jute-growing period, one hectare of jute plants can absorb about 15 metric tons of $CO_2$ from the atmosphere and release about 11 metric tons of oxygen.[3]
- In 100 percent form, jute fabric is biodegradable and can be broken down into simpler substances by microorganisms, light, air, or water in a nontoxic process. Absolute biodegradability depends on the dyes and trims used, and route of disposal.

  ♦ See **Appendix: Biodegradability in the Fashion Fibers STUDIO** for more design guidance.

- Due to its extensive root system, jute can help reduce soil loss and erosion and is particularly suitable for crop rotation. Since the leaves of the plant are left in the field after harvest, the nitrogen they contain absorbs into the soil, and food crops can be grown immediately without having to leave the fields fallow.
- Once the jute fiber is extracted from the stem, processing it into yarn is largely mechanical with minimal environmental impact.

# Potential Impacts

### Water and Chemicals Used for Extracting Fiber from Plant

Jute is a bast fiber and is extracted directly from the stalk of the plant in a process similar to that used for flax, hemp, and bamboo (for linen). The fiber is extracted through a process called retting, which separates the fiber from the stems using microorganisms and moisture. This is carried out in the field (as with dew retting) or in tanks (in water or chemical retting). Dew retting is preferred, as it utilizes the natural moisture of dew, but is the longest process, taking over two to three weeks to break down the stems. Although chemical retting is the fastest process, the wastewater is concentrated and rich in chemicals and biological matter and must be treated before its release; otherwise it can harm aquatic ecosystems, receiving water bodies, and drinking water sources.[4] See Part 1 Table 2 for more detailed information about the advantages and impacts of each type of retting.

### Dyeing and Treatments

The natural color of jute fiber is beige, and jute yarn or fabric must be bleached with chlorine to render it light enough to receive dyes for light or clear shades. Chlorine bleach can form halogenated organic compounds in the wastewater. These compounds bioaccumulate in the food chain, are known teratogens (can cause birth defects) and mutagens (can change genetic material), are suspected human carcinogens (cancer causing), and cause reproductive harm.

♦ Visit **Chapter 22: Bleaching** for information on the environmental impacts of bleaching.

**Table 5.1 Optimize the Sustainability Benefits of Jute**

| Opportunity | Benefit | Considerations |
| --- | --- | --- |
| Promote suppliers using organic jute. | • In addition to the general ecological benefits of jute, organic processes ensure that no disallowed pesticides or fertilizers are used. | • Organic certification must be in place by a recognized international certification agency accredited by **IFOAM**.<br>• Although organic certification disallows the use of chemicals in the growing of jute, it does not necessarily guarantee low water use, fair labor practices, or a fair price to the farmer.<br>• Organic jute is not as readily available as conventional jute, and it commands a premium. |
| Promote the use of natural color jute. | • No bleaches or dyes are used in this case, and associated pollution impacts are avoided. | • Using the natural color of jute will limit your color palette to neutrals. Spice up a neutral color palette with interesting drape, silhouettes, or seaming. |
| Promote suppliers who use dew retting over water or chemical retting. | • Dew retting reduces the biological load in the receiving water bodies and adds nutrients to the soil. | • The natural color may vary slightly from lot to lot, since the process is influenced by weather. |
| Promote suppliers who use enzymatic retting—an alternative to water or chemical retting. | • The process is faster and leaves the water unharmed. Enzymatic retting can be commercially reproduced. | • Low fiber strength. The process is less common than other retting processes. |
| Use hydrogen peroxide to lighten the natural beige color for dyeing dark shades and bright/light shades.<br>♦ Visit **Chapter 22: Bleaching** for more guidance. | • Hydrogen peroxide harmlessly decomposes into water and oxygen gas. | • Non-chlorine bleaches do not strip out the original color of the fiber. Consequently, colors will be duller due to the overdyed effect. Non-chlorine bleaching is adequate for dark colors, which mask the original beige tone. |
| Use ozone bleaching processes, which use ozone gas (a much more powerful oxidizing agent than chlorine) to strip out the natural beige color of jute.<br>Promote the particular aesthetic of ozone bleach effects. | • It is very effective at fading pre-dyed/preprinted fabrics and garments, can be performed at lower temperatures than with hydrogen peroxide, and uses no water at all. In addition, ozone is completely biodegradable, as it reverts rapidly back to oxygen ($O_2$), leaving no chemical residue. | • Ozone has limited availability and is relatively expensive since it requires investment in ozone-generating equipment.<br>• Ozone processes produce a different aesthetic than chlorine derivative or permanganate bleaching. |
| Promote the use of enzymes to strip out the natural beige color of jute.<br>Promote the particular aesthetic of enzyme bleaches. | • Enzymes use less water than bleaching.<br>• Enzymes are fully biodegradable.<br>• Enzymes are gentler to the fiber than bleaching, and do not cause degradation as bleaching could.<br>• The wastewater for enzymes is easier and cheaper to treat than bleaching. | • Enzymes are not allowed under **Global Organic Textile Standard (GOTS)** standards because most technical enzymes are produced with genetically modified organisms (GMOs). GMOs and enzymes are not allowed for use in products labeled as "organic" or "made with organic," according to GOTS standard.<br>• Enzymes produce a different aesthetic than chlorine derivative or permanganate bleaching. |
| Promote the use of jute-blended fabrics. | • Can achieve the property benefits of both fibers. | • Sometimes requires further processing, which could include chemicals. |
| Know the difference between natural jute fabric and jute made from a viscose process. | • Both reduce deforestation from fibers sourced from wood trees. | • Jute viscose is chemically processed and has greater pollution impacts to water and air.<br>♦ Visit **Chapter 17: Rayon/Viscose Made from Wood** for more guidance. |

Jute can also be blended with wool. By treating jute with caustic soda (also called "lye"), crimp, softness, pliability, and appearance are improved, aiding in its ability to be spun with wool. Due to its toxic nature, even a small quantity of caustic soda in a diluted solution can severely injure the eyes, cause blindness, or cause skins burns. Personal protective equipment is required for workers handling caustic soda, although that is not always enforced in factories.[5]

Liquid ammonia has a similar effect on jute, improving crimp, softness, pliability, and appearance, as well as the added characteristic of improving flame resistance when treated with flameproofing agent.[6] Exposure to high concentrations of ammonia in air causes immediate burning of the eyes, nose, throat, and respiratory tract and can result in blindness, lung damage, or death.[7] Again, personal protective equipment is required to handle liquid ammonia, but this is not always enforced in factories.

### Consumer Care/Washing

Jute may be machine washed or dry-cleaned. Since the fabric wrinkles easily, it often requires heavy pressing to render it smooth after wash. To iron a shirt for fifteen minutes once every other week would require 7.15 kWh annual energy consumption, which at 11 cents per kWh is about 79 cents per year.[8] While it is important to know that a jute garment will require heavy pressing and electricity use in its lifetime, the amount of electricity is relatively low, and it might be useful to instead address other potential impacts that could have more influence.

- Visit **Appendix B: Consumer Care and Washing** for more information.
- Visit **Chapter 23: Dyeing and Printing** for more information on impacts associated with dyeing.
- Visit **Chapter 26: Recycled/Circular Textiles Technologies** for more information about recyclability.
- Visit **Appendix: Biodegradability in the Fashion Fibers STUDIO** for information about designing for biodegradability.

## Availability

Jute is readily available in 100 percent as well as blends with wool and silk.

About 95 percent of the world's jute is grown in India and Bangladesh. Nepal, Myanmar, China, Thailand, Vietnam, and Brazil also produce jute.[9] Pakistan, although it does not produce much, imports a substantial amount of raw jute, mainly from Bangladesh, for processing.[10]

A number of farmers in Bangladesh are currently growing organic jute. Organic certification by an internationally recognized certification agency accredited by IFOAM must be in place.

## Fashion Applications

As 100 percent, jute is highly durable and ideal for medium- to heavyweight blazers, coats, or jackets. Jute may also be used for lightweight to medium-weight trousers.

**BOX 5.2**
**FASHION APPLICATIONS**

As 100 percent, jute is highly durable and ideal for medium- to heavyweight blazers, coats, or jackets.

**Figure 5.2**
Jute jacket and pants.
*Stefania D'Alessandro/Getty Images*

# Potential Marketing Opportunities

- **Fast-growing natural resource:** Jute is considered a fast-growing resource, as seen in Part 1 Table 1, and can be marketed as such.
- **Low water footprint in growing:** Jute is biologically efficient, mainly rain fed, and can grow in diverse climates.
- **Biodegradable:** All fibers, yarns, trims, and dyes used to manufacture the product or garment must also be biodegradable or disassembled before disposal. This should be substantiated with documentation that the product can completely break down into nontoxic material by being processed in a facility where compost is accepted. A secondary label or marketing material should be provided to instruct the customer.

- **Non-chlorine bleached:** If alternative bleach is used. Specify the alternative bleaching method and explain the environmental benefits on the secondary label or marketing material.
- **Organic:** If organic jute fiber is used. All fibers, yarns, trims, and dyes used to manufacture the garment must comply with the GOTS organic garment standard. Simply state "made from 100 percent organic jute," if this is verified and accurate.
- **Dew retted:** If dew retted processed.
- **Enzyme retted:** If enzyme retted processed.
- **Renewable resource:** Jute is a renewable resource and can be replaced or replenished by natural processes.

# INNOVATION EXERCISES

## Design

1. Design a garment that emphasizes one or more of jute's sustainability benefits.
2. Design a garment that circumvents one or more of jute's environmental impacts.
3. Design a "culturally durable" (i.e., styling that doesn't date with passing trends) garment in jute to optimize the fiber's physical durability.
4. Design footwear or accessories from jute to emphasize its natural durability.
5. Design a garment that uses jute fiber in blends with cotton to negate the need for using chlorine bleach.
6. Design a garment where jute is strategically placed in areas of high stress to maximize its physically durable properties.
7. Design a pair of 100 percent jute jeans. Explain the benefits.
8. Develop a collection that is entirely designed to be easily deconstructed to enable a take-back and recycling program. Experiment with seaming and a variety of disassembly mechanisms in different fabrics.

## Merchandising

9. Create a fifty-word marketing communication message for consumers about the sustainability benefits of jute.
10. Incorporating information from chapters 1, 2, 3, 4, and 5, develop a table that a retail buyer could use to quickly compare the advantages and disadvantages of 100 percent conventional cotton, 100 percent organic cotton, 100 percent flax, 100 percent bamboo linen, 100 percent hemp, and 100 percent jute.
11. Create a retail point-of-purchase sign or hangtag to promote fashion merchandise made from 100 percent jute.
12. Go online and find three fashion retailers who sell fashion merchandise made from 100 percent jute. What are the retailers and fashion merchandise offered? Why did they select this fiber for their fashion merchandise? How do they promote the advantages of this fiber to consumers? Do you believe their promotional strategies are accurate and effective? Why or why not?
13. Develop a marketing research protocol (e.g., focus group protocol, online survey, observational strategy) to determine consumers' perceptions of jute for fashion merchandise.

## Innovation Exercise Example

**Design a garment where jute is strategically placed in areas of high stress to maximize its physically durable properties.**

**Figure 5.3**
This design shows a jute/cotton children's playsuit. In this example, the body is 100 percent medium-weight cotton to allow movement and 100 percent jute in the pockets, knee pads, and shoulders to support durability in areas of high stress.
*Amy Williams*

# REVIEW AND RECOMMENDATION

## Preferred

Jute fiber is preferred for use as a sustainable fiber because:

- Jute is a fast-growing, renewable fiber that is annually farmed.
- Jute is a biologically efficient, low-maintenance crop that requires few chemical inputs during the growing season. It is mainly rain fed, traditionally farmed, and grown similarly to organic produce.[11]
- Jute is biodegradable.
- Jute can reduce soil loss and erosion and is particularly suitable for crop rotation.
- Studies reveal that the $CO_2$ assimilation rate of jute is several times higher than that of trees.
- Once the jute fiber is extracted from the stem, processing it into yarn is largely mechanical with minimal environmental impact.

## Recommended Improvements

- Jute fiber that has been either dew retted or enzyme retted, as it reduces the biological load in the receiving water bodies
- Jute fiber that has been bleached with alternative methods to prepare the fiber for dyeing, such as ozone bleaching or hydrogen peroxide or enzyme processes
- Incorporating environmental dyeing practices
- Developing end-of-use strategy (whether compostability or recyclability)

Tim Graham/Getty Images

# Wool

The term "wool" means the fiber from the fleece of the sheep or lamb or hair of the Angora or Cashmere goat (and may include the so-called specialty fibers from the hair of the camel, alpaca, llama, and vicuna) that has never been reclaimed from any woven or felted wool product.[1] This chapter will focus primarily on wool from sheep, including the better-known merino.

Merino sheep are just one of hundreds of different breeds of sheep. Merino produce the majority of apparel wool, going into clothing and fashion, globally. The merino breed is best suited to produce apparel wool due to its unique fiber characteristics, which include lower fiber micron diameter, making finer yarns and fabrics; softer "handle," making it softer to wear next to skin; and the white color, making it easier to dye uniformly.

Wool is produced in highly diverse sheep farming systems and has a wide range of fiber length, diameter, and strength characteristics.

**Figure 6.1**
Merino wool is naturally thermoregulating due to its natural crimp.
*kycstudio /Istock Photo*

# Sustainable Benefits

- Wool is a natural fiber and renewable.
- Sheep contribute in many areas to maintain and regenerate grasslands and landscapes, aiding in biodiversity and soil improvement.
- Wool is thermoregulating and can be worn all year. Some wool fibers, such as merino, have a high crimp, which allows for air to fill in the space in the fiber's gaps. This air serves as an insulation layer, in both summer and winter, against heat and cold.
- Wool can absorb more than 35 percent of its own weight in moisture without feeling wet to the touch. If the ambient air is warm, the moisture dries faster, resulting in a refreshing evaporative coolness.
- Wool is naturally water repellent, with the degree dependent on how coarse the wool is and the density of the fabric and its surface. Loden and felted wools are particularly water repellent.[2]
- The surface of wool fiber is to a certain degree dirt and stain repellant, but once a stain or dirt has been absorbed, it sticks and is harder to remove than from other fibers.
- Wool is naturally **fire resistant** due to its high nitrogen and water content.[3,4] There is no need to use fire-resistant coatings or synthetic topical finishes on wool; it may be used as a viable (and better!) alternative to synthetic fabrics with **fire-retardant** treatment. A loose weave or knit, which allows more airflow, will lower wool's ability to be fire retardant.
- Wool absorbs odors due to the protein molecules and is, to some extent, self-cleaning. Wool tends not to smell bad, even after long periods of wear. Because of this, wool garments do not need to be washed frequently—just air out after use—and are great for sports- and workwear.
- Washing of wool is generally done at low temperatures, which saves energy.
- One hundred percent wool is biodegradable and can be broken down into simpler substances by microorganisms, light, air, or water in a nontoxic process. Absolute biodegradability depends on the dyes and trims used, as well as route of disposal.

**Table 6.1 Wool Fiber Characteristics**

| | Merino Wool | Coarse Wool |
|---|---|---|
| Microns | • 10–24.5 microns | • 28–33 microns |
| Major producers | • Australia, New Zealand, South Africa, Argentina, and Uruguay | • Turkey, India, Russia, United States, New Zealand, and United Kingdom |
| Fiber collection | • Shearing | • Shearing |
| Blends well with | • Natural and synthetic fibers | • Natural and synthetic fibers |
| Application | • Next-to-skin garments like underwear, babywear, sleepwear, and base layers<br>• Knitwear, woven suits, and fashion products<br>• Outerwear such as coats, gloves, hats, and scarves<br>• Durable blankets and bedding | • Knitwear, outerwear, accessories, rugs, carpets, and interior textiles |
| Natural colors | • Off-white | • White, brown, gray, charcoal, and black |

# Potential Impacts

## Animal Welfare

Wool farmers have focused on breeding of merino over the years to provide those unique merino wool fiber qualities that are sought after—low micron fiber diameter, soft handle, and white wool.

Merino sheep were bred to have wrinkles, as there is a modest positive genetic correlation between higher fleece weight and higher wrinkle. New technology (Sheep Breeding Values) is making it easier to find low-wrinkle, high-fleece-weight sheep that also tend to produce more lambs. As a direct consequence, more and more farmers have been breeding "plainer" merino sheep in recent decades with fewer wrinkles.

A key reason farmers have been focused on breeding merinos with fewer wrinkles is a serious health condition called fly-strike. Fly-strike occurs during warm, wet weather when female flies are searching for a place to lay their eggs. The female flies then lay their eggs in the long breech wool in a sheep's rump where fecal and urine stain. All sheep breeds are susceptible to getting fly-strike. Fly-strike is exacerbated when a certain fly, *Lucilia cuprina,* is present. This fly is most prevalent in Australia with its hot, wet summers; merino sheep are more likely to suffer from fly-strike (the eggs hatch into larvae that feed on the moist, weakened skin and underlying tissue) due to their dense wool, which stays moist for longer periods. If left untreated, sheep will die a long and painful death due to blood poisoning.

Increasingly there are more treatments available to prevent fly-strike. However, the most common is the mulesing procedure—a surgical procedure conducted by trained individuals in which small pieces of excess skin are removed from the breech area of the sheep. The aim of this procedure is to create a smoother breech area that is less prone to flies and hence providing a lifelong preventative treatment against fly-strike.

Mulesing has become a heavily discussed subject among activists groups and the textile industry as a whole. Some activist groups claim this procedure constitutes cruelty to animals, and there have been brand boycotts and countrywide phaseout plans of mulesing. Research reflected in the positions of leading welfare groups, RSPCA Australia and Australian Veterinary Association, points to mulesing as the most effective way to prevent fly-strike. There is evidence that some pain caused at mulesing can be reduced by applying topical anesthetic treatment after the procedure.

Considerable research has been conducted over the years to find alternatives to mulesing with some success, producing a range of breeding tools to assist farmers: breech clips, new chemical treatment, and intradermal technology all designed to safeguard the merino sheep from suffering from fly-strike.

Some sheep can be treated with chemicals to control lice (another external parasite in the wool that causes the sheep to rub) and internal parasites (intestinal worms). These chemicals may be applied directly to the fleece or by submerging the sheep into chemical solution pools (sheep dips) or being treated with a backline chemical post-shearing. The damage from internal parasites varies among climatic regions, but in the higher rainfall areas where there are higher worm burdens on the pastures, sheep need to be drenched to keep them fit and healthy.

However, in Australia there was a significant reduction in organophosphate and synthetic pyrethroid residues, with levels falling from approximately 7 percent in the 1990s to approximately 1 percent in the early 2000s.

### Land Use

Historically, intensive sheep ranching has contributed to land degradation, or desertification, in some regions. Desertification happens when available rainfall becomes noneffective (it disappears through flooding and evaporation). However, it is important to note that there are increasingly modern holistic sheep- and cattle-grazing systems that have been shown to benefit biodiversity and ecosystem health and restore degraded grasslands. In such methods, sheep can contribute to land regeneration when using holistic grazing methods like "Allan Savory" in areas with deserted grasslands (e.g., in Patagonia). In that case, wool is restorative beyond sustainability.

### Chemical Use for Processing Wool from Sheep Fiber

Raw wool contains around one-third grease or lanolin, which is often recovered. To remove these substances from the wool fiber, a cleaning or **scouring** process is carried out at hot temperatures (approximately 60–65°C) in an aqueous solution of sodium hydroxide, sodium carbonate, or sodium sulphate, and detergent.[5] Scouring consumes large amounts of water and produces an effluent with high **biological oxygen demand (BOD)** and high suspended solids content.[6] Several technologies have been developed to reduce the BOD of effluent from wool scours, which is controlled by strict government regulations in the major processing countries.

High BOD reduces the **dissolved oxygen (DO)** levels, meaning less oxygen is available to fish and other aquatic organisms. Trace amounts of pesticides are normally removed from the effluent before it is released to the environment. Some of these detergents used for scouring are banned in Europe, but not elsewhere. State-of-the-art scouring plants around the world are following strict regulations and best practices and often have closed systems for reuse of water and collection of sludge and grease.

### Shrink-Proofing Treatments

Anti-shrinking treatments prevent wool from felting during wash and generally have included chlorine in some form. Chlorine-Hercosett is a treatment used on wool tops and fiber. Repeated exposure to chlorine can affect the human respiratory system. The wastewater from the wool chlorination process contains chemicals of environmental concern. Recent developments like Schoellers EXP and Südwolle use environmentally friendly plasma treatments instead.

### Dyeing

Wool, like all other cellulosic fibers, can be dyed with either natural or synthetic dyes, and both have their considerations. Certain types of synthetic dyes are suspected human carcinogens (cancer causing) and mutagens (can change genetic material), and dye water can negatively impact receiving water bodies, harm aquatic ecosystems, and affect local drinking water if left untreated before its release. Natural dyes require a **mordant** (to achieve color fastness) and heavy metal salts, which should also not be left untreated before the wastewater is released to water bodies. Natural dyed fiber can also be a challenge when it comes to consistency of color, colorfastness, lightfastness, and durability.

Wool can also be used without dyes in natural pigment or in various shades (coarse wool).

- Visit **Chapter 23: Dyeing and Printing** for more information on impacts associated with dyeing.

### Consumer Care/Washing

Wool needs to be laundered less often due to its natural ability to repel stains and odors. Simply setting it to air overnight can bypass the need to wash. While machine washing on the delicate cycle and spin cycle is not only safe but also recommended, some wool garments are labeled "dry-clean only," which can cause potentially toxic dry-cleaning chemicals to be used unnecessarily.

While vigorous movement (e.g., hand washing) can cause wool to felt, your machine's spin cycle is recommended, as the clothing "sits still" inside the machine and allows dirty water to drain.

For suits, dry-cleaning is seldom necessary provided spills are wiped off immediately. Wool suits can be aired or can take a steam bath in your bathroom.

Wool should not be washed with commercial detergents, avoiding environmental impacts associated with commercial detergents on the market. Conventional detergents use enzymes to clean the dirt, and enzymes eat wool (proteins).

It is recommended to use shampoos or detergents such as Woolite.

- Visit **Appendix B: Consumer Care and Washing** for more information.

## Reuse, Biodegradability, and Recyclability

### Reuse

Wool is one of the most reused fibers, having been found to account for up to 5 percent by weight of the total clothing donated by consumers for recycling and reuse, substantially higher than wool's share of the virgin fiber supply of about 1.3 percent.[7] Typically three-quarters of these donated clothes are reused directly through charities, with their second use having no environmental effect from fiber production and manufacturing.

- Visit **Appendix: Biodegradability in the Fashion Fibers STUDIO** for information about designing for biodegradability.

### Recyclability

Wool is **recyclable**, and there are many suppliers producing recycled wool yarns that are used for apparel. The fiber is mechanically recycled, which means it mechanically is pulled back into raw fiber, through chopping and carding of the fiber, and then spun back into a yarn.

Wool recycling has a long history dating back to the activities of Benjamin Law in Bradford, UK, in 1813. About one-quarter of all donated clothes are recycled through open- and closed-loop paths that are commercially well established and extend the active life of the wool fiber by many years beyond the first use.

**BOX 6.1**
**WOOL IS COMMERCIALLY RECYCLABLE**

Recycled wool and recycled cotton are the two most often used recycled fibers in the fashion industry. Recycled wool is suitable for all fashion applications that virgin wool is.

**Figure 6.2**
Suit made with recycled wool by Swedish brand Uniforms for the Dedicated, uniformsforthededicated.com.
*Uniforms for the Dedicated*

When considering a fiber's ability to be recyclable, it is necessary to look at whether it can be "**upcycled**" or "**downcycled**." To upcycle means to be able to create a yarn or product that is of equal or greater value than the original. This is always the goal, because downcycled materials eventually end up in the trash, sometimes after a very short period of use. Since the mechanical recycling process would shorten the fiber length, the fiber would need to be blended with virgin fiber (or another natural fiber) to help facilitate the processing and strengthen the yarn. Eventually, though, the staple fiber would get too short and not be strong enough for apparel or other use and would end up in the trash.

Due to its natural flame resistance, wool fiber is often recycled in nonwoven applications, such as mattress insulator pads, where the wool content contributes to the fire retardancy of the product. This type of application can extend the use of the wool fiber to seven years and more.

- Visit **Chapter 26: Recycled/Circular Textiles Technologies** for more information about recyclability.

**Table 6.2 Optimize the Sustainability Benefits of Wool**

| Opportunity | Benefit | Considerations |
|---|---|---|
| Promote charities that collect and resell clothing. | • Wool is one of the most highly donated fiber types. | • Reused clothing has a very light environmental footprint, involving no fiber production and processing. |
| Promote the use of nonmulesed merino Australian wool. | • Ensures that ethical treatment of animals is taking place. | • Work with Australian Wool Innovation Limited (AWI) or e-wool, which can provide a list of Australian merino ranchers that have incorporated methods to replace mulesing.<br>• Visiting a ranch where your wool is sourced from is the best way to ensure animal-friendly practices are being implemented. Your company might purchase either yarn or the fabric, so it will be difficult to find out where the fiber has been sourced from (difficult, but not impossible).<br>• Visits require travel budgets. Build these visits into your annual budget.<br>• The Textile Exchange is establishing a standard to address animal welfare and land management of sheep. The Responsible Wool Standard launched in 2016 and will help designers and brands ensure they are purchasing responsible wool. |
| Promote suppliers using recycled wool. | • Available in Northern England and Prato, Italy. Since wool is a renewable resource, the primary benefit of recycled wool is that it is reducing the use of virgin resources and the environmental impact associated with processing and production of virgin wool. Using recycled wool may also ease the pressure that industrialized sheep ranching places on the land. | • Recycling wool creates shorter fibers and a weaker yarn, which is why it needs to be blended with a percentage of virgin wool or synthetic fiber to maintain strength for finer-count yarns. The coarser the yarn count, the less virgin wool or synthetic fiber is required. |
| Promote suppliers using Cardato Regenerated $CO_2$ Neutral products.[8] | • The Cardato Regenerated $CO_2$ Neutral brand certifies both the carbon footprint of the textile production process and the use of regenerated raw materials. To carry the label, products must be produced in Prato, Italy; be produced with at least 70 percent of recycled material (recycled clothing or textile off-cuts); made by mills that have accounted for their $CO_2$ emissions; and the company must have purchased emission credits from the Prato Chamber of Commerce. | • Could be more expensive than traditional wool. Will need to verify with suppliers. |
| Promote farmers who are using regenerative farming practices such as Allan Savory. | • Farming practices aim to improve biodiversity, revert land degradation, and promote human treatment of animals. | • Ensure that the ranches you're working with promote these practices. |
| Ensure that you are using a supplier that treats the effluent after the scouring process. | • Less biological load and toxic chemicals in the wastewater. | • Usually wool is purchased as a yarn or a fabric, but scouring occurs deeper in the supply chain. Finding out where your wool is sourced from might take some time and effort. |

*(continued)*

**Table 6.2** *(continued)*

| Opportunity | Benefit | Considerations |
|---|---|---|
| Promote the use of chlorine-free or AOX-free wool. | • Chlorine is not used during the shrink-proofing process. | • Not as readily available and could cost more.<br>• Usually wool is purchased as a yarn or a fabric, but scouring occurs deeper in the supply chain. Finding out where your wool is sourced from might take some time and effort. |
| Promote wool products with the OEKO-TEX® certified label.[9] | • Ensure that products pose no risk to health. These products do not contain allergenic dyestuffs or dyestuffs that form carcinogenic aryl-amines and several other banned chemicals. The certification process includes thorough testing for a long list of chemicals.<br>♦ Visit **Chapter 23: Dyeing and Printing** for more information on impacts associated with dyeing. | • Not as readily available and could cost more.<br>• Go to www.oeko-tex.com to find suppliers of OEKO-TEX® certified wool. |
| Promote wildlife-friendly grazing practices for Kashmir goats.<br>Prioritize sites that have endangered wild species. | • Decreases impacts of overgrazing and loss of biodiversity due to desertification. | • Will require collaboration with nongovernmental organizations (NGOs) and industry partners. |
| Develop relationships with producers and monitor farmers. | • Ensures animal-friendly practices are being implemented. | • Again, since wool is generally purchased as a yarn or a fabric, it might be difficult to identify where your wool is being sourced from.<br>• Rely on Textile Exchange's Responsible Wool Standard to ensure that humane practices are taking place. |

# Availability

Wool garments are collected for reuse in many countries. Recycled wools are readily available in the West Yorkshire, UK, and Prato, Italy.

Non-mulesed merino wool is available in Argentina, Uruguay, New Zealand, South Africa, and even in certain areas of Australia, since the blowfly does not exist in these areas.

# Fashion Applications

The applications for wool vary according to the type of fiber and breed of sheep and animal, but generally include both woven and knit applications from light- to heavyweights.

Recycled wool lends itself more to knit sweaters and coarser fabrics, though smaller percentages of recycled wool are found in high-end tweed fabrics made by Italian mills.

## Potential Marketing Opportunities

- **X percent post-consumer recycled wool (if used):** Wool from post-consumer channels, such as from collected and unused clothing.
- **X percent post-industrial recycled wool (if used):** Wool from post-industrial recycled channels, such as trimmings and cuttings from the industrial process.
- **Non-mulesed wool:** Must ensure that sheep have not been mulesed.
- **Biodegradable:** All fibers, yarns, trims, and dyes used to manufacture the product or garment must also be biodegradable or disassembled before disposal. This should be substantiated with documentation that the product can completely break down into nontoxic material by being processed in a facility where compost is accepted. A secondary label or marketing material should be provided to instruct the customer.
- **Needs less washing:** Wool needs to be laundered less often due to its natural ability to repel stains and odors. Inform your customers that in some cases they can bypass the need to wash by setting the garment out to air overnight.

# INNOVATION EXERCISES

### Design

1. Design a garment that emphasizes one or more of the sustainability benefits of wool.
2. Design a garment that circumvents one or more of the environmental impacts of wool.
3. Design a garment using naturally pigmented wool in an innovative way.
4. Create a garment using both wool and cotton where you agitate in hot water to felt the wool and pucker the cotton.
5. Design a garment that uses wool in strategic areas to emphasize its self-cleaning and moisture-absorbent attributes.
6. Go to your local fabric or yarn store and find yarn or fabric made from 100 percent wool fiber. Identify the source of the fiber, including name of ranch and location. Plan a visit if you live close!
7. Design a garment that utilizes the best features of wool, like comfort and climate, in layer-by-layer, next-to-skin, or wool for summer.

### Merchandising

8. Create a fifty-word marketing communication message for consumers about the sustainability benefits of wool.
9. Incorporating information from chapters 1, 2, 3, 4, 5, and 6, develop a table that a retail buyer could use to quickly compare the advantages and disadvantages of 100 percent conventional cotton, 100 percent organic cotton, 100 percent flax, 100 percent bamboo linen, 100 percent hemp, 100 percent jute, and 100 percent wool.
10. Create a retail point-of-purchase sign or hangtag to promote fashion merchandise made from 100 percent wool.
11. Go online and find three fashion retailers who sell fashion merchandise made from 100 percent wool. What are the retailers and fashion merchandise offered? Why did they select this fiber for their fashion merchandise?

## Innovation Exercise Example

**Design a garment that uses wool in strategic areas to emphasize its self-cleaning and moisture-absorbent attributes.**

**Figure 6.3**
In this design example, wool is used as an inset in areas where one would most often perspire—under the arms and the back of the legs.
*Amy Williams*

How do they promote the advantages of this fiber to consumers? Do you believe their promotional strategies are accurate and effective? Why or why not?

**12.** Develop a marketing research protocol (e.g., focus group protocol, online survey, observational strategy) to determine consumers' perceptions of wool for fashion merchandise.

# REVIEW AND RECOMMENDATION

## Not Preferred

- Wool from sheep that has been produced with inhumane treatment
- Wool that has been treated with chlorine anti-shrink treatments

## Preferred

- Wool from sheep that have been treated with good animal welfare practices and certifications/documentation
- Wool that has been scoured with alternative processes to promote clean water effluent and closed-loop processes

## Recommended Improvements

- Work with ranchers to combat fly-strike through holistic means, and use mulesing and chemical applications as a last resort
- Incorporate environmental dyeing practices
- Develop end-of-use strategy (whether compostability or recyclability)

Kraivuttinun/iStock

# 7

# Silk

Silk is a protein fiber produced by silkworms. As a silkworm develops into an adult, it feeds on leaves and then spins a cocoon from one continuous silk strand or filament, approximately 1,000 yards long.[1] Silkworms have a life cycle much like caterpillars and butterflies: inside the cocoon the worm changes into a chrysalis, then into a moth, which then seeks to leave the chrysalis.[2]

Heat is used to soften the hardened filaments so they can be unwound. Single filaments are then combined with a slight twist into one strand, a process known as filature or "silk reeling."

# Sustainable Benefits

- Silk is a renewable natural resource. Since the silk filament is a continuous thread, it has great tensile strength. In woven fabrics, silk's triangular structure acts as a prism that refracts light, giving silk cloth its highly prized "natural shimmer."
- Silk has good absorbency and low conductivity, and it dyes easily.
- In 100 percent form, silk fiber is biodegradable and can be broken down into simpler substances by microorganisms, light, air, or water in a nontoxic process. Absolute biodegradability depends on the dyes and trims used, as well as route of disposal.
- Sericulture (silk farming) is labor intensive. About one million workers are employed in the silk sector in China. Sericulture provides income for 700,000 households in India and 20,000 weaving families in Thailand. Wild silk can provide a year-round income for tribal people in India and some areas of China.[3]

# Potential Impacts

### Animal Welfare

On domesticated silk farms, the chrysalis is killed to prevent the moth from making a hole in the cocoon. The reason for this is that the hole breaks the highly prized long silk filament into thousands of short lengths, which are useless for higher-quality spinning.

**Figure 7.1**
Woman putting boiled silk cocoons into a reeler.
*fototrav/iStock*

**BOX 7.1 SUSTAINABLE BENEFIT**

About one million workers are employed in the silk sector in China. Sericulture provides income for 700,000 households in India and 20,000 weaving families in Thailand.[4]

### Processing

Cocoons are soaked in sodium carbonate to soften them in preparation for reeling (unwinding the filament from the cocoon). Silk fabric is then woven with the natural gum, or **sericin**, still on the yarn, acting as a natural **sizing agent**. After weaving, the gum is removed by boiling the fabric in alkali. This can result in a 20 percent reduction of the harvested weight of the silk. Some of this lost weight is added back by saturating the silk fabric in a bath of tin-phosphate-silicate salts. These processes can create a high biological load on the water and deplete available oxygen for aquatic species if left untreated. Exposure to tin through breathing and skin contact can have acute and long-term effects on worker health if proper equipment is not used.[5]

Lightweight silk fabrics (fine gauge silk) are prone to wear and are degraded by exposure to sunlight and hot temperatures. They can also be susceptible to abrasion and twisting in laundering.[6]

**Table 7.1 Optimize the Sustainability Benefits of Silk**

| Opportunity | Benefit | Considerations |
|---|---|---|
| Promote the use of wild, or "Tussah," silk. | • Wild, or "Tussah," or "tasar," silkworms feed on leaves, not necessarily mulberry, and do not harm the chrysalis. Tussah silk is derived from cocoons collected after the moth has emerged naturally in the field.<br>• Wild silk provides a year-round income for tribal people in India and some areas of China.[7] | • Because the continuous silk fiber is broken into smaller pieces as the moth leaves the cocoon, wild silk has a rougher and slubbier surface than cultivated silk.<br>• Due to the shorter (less prized) fiber length, wild/Tussah silk is less expensive than domesticated silk.<br>• Tussah silk fabrics have a coarser texture and are typically stiffer and heavier than domesticated silk.<br>• Wild/Tussah silk is available in small quantities. |
| Promote the use of ahimsa silk.[8] | • This is cultivated in India and doesn't require the chrysalis to be killed.<br>• The fibers are spun into "slubby threads" instead of reeled.<br>• Ahimsa silk is softer and finer than regular silk and has a pearl matte natural finish. | • It is more costly than regular silk due to its laborious process of spinning the many pieces of yarn into one continuous thread.<br>• Not all slubby silks are ahimsa silk.<br>• Manufacturers often label these slubby silks as Dupioni or shantung, and claim they are ahimsa silk. This should be substantiated with documentation. |
| Promote the use of **organic** silk. | • Since pesticides are rarely used on silk fiber production (this would kill the silkworm), the main benefit of organic certification is using organically cultivated mulberry bushes.<br>• Organic cultivation has wide-ranging benefits for the surrounding ecosystem. | • Organic silk is available in small quantities and carries a price premium.<br>• Certification of organic silk must be in place by an internationally recognized certification agency accredited by the **International Federation of Agriculture Movements (IFOAM)**. |
| Promote the use of **Fair Trade** silk. | • Ensures the proper treatment of workers. | • Fair Trade silk products are less available than conventional silk. It does not necessarily mean "organic." |
| Blend silk with organic cotton, organic wool, organic linen, etc. | • Brings a "luxury" element to the product and commands a higher retail price.<br>• Blending with a washable fiber reduces the impact of consumer care/dry-cleaning. | • When developing silk/organic fiber blend products, make sure to take type of dyes used into consideration too, as they can have a significant impact on the complete environmental story. |

**BOX 7.2 PROMOTE THE USE OF ORGANIC SILK**

While pesticides are rarely used in silk fiber production (this would kill the silkworm), organic certification is using organically cultivated mulberry bushes, which can have wide-ranging benefits for the surrounding ecosystem.

**Figure 7.2**
Dress made from organic silk and naturally dyed with tea. Made by Averti. www.averti.la.
*www.averti.la*

### Consumer Care/Washing

Due to the delicacy of the fabric, silk products are typically hand washed or dry-cleaned. Silk tends to crush and wrinkle more easily. This wrinkling creates a need to increase the frequency of ironing. Electricity and chemicals for washing must be considered when weighing out sustainability benefits of a fiber.

- Visit **Chapter 23: Dyeing and Printing** for more information on impacts associated with dyeing.
- Visit **Chapter 26: Recycled/Circular Textiles Technologies** for more information about recyclability.
- Visit **Appendix B: Consumer Care and Washing** for more information.
- Visit **Appendix: Biodegradability in the Fashion Fibers STUDIO** for information about designing for biodegradability.

# Availability

China produces about 70 percent of the world's silk, followed by Brazil, India, Thailand, and Vietnam, with minor production in Turkmenistan and Uzbekistan. India, Italy, and Japan are the main importers of raw silk for processing.[9]

Organic silk is available in small quantities at premium prices. Certification of organic silk must be in place by an internationally recognized certification agency accredited by IFOAM.

Most wild silk is cultivated in China, India, and Japan.[10]

Ahimsa silk is cultivated in India. Ahimsa and peace silk allow the mulberry-fed moth to leave the cocoon before it is harvested.

# Fashion Applications

One hundred percent silk is available in a variety of fabrics and numerous blends with other fibers, such as cotton and wool.

Silk's natural beauty and other properties—such as comfort in warm weather and warmth during colder months—have made it sought after for use in high-fashion clothes, lingerie, and underwear.[11]

Due to its coarseness, wild silk is largely used in furnishings and interiors.[12]

# Potential Marketing Opportunities

- **Fair Trade/artisan wild silk:** When developed through a nonprofit organization and source is verified.
- **Wild silk:** With verification of source in place.
- **Ahimsa silk:** With verification of source in place.
- **Renewable natural resource:** Silk is considered a renewable natural resource and can be replaced or replenished by a natural process.

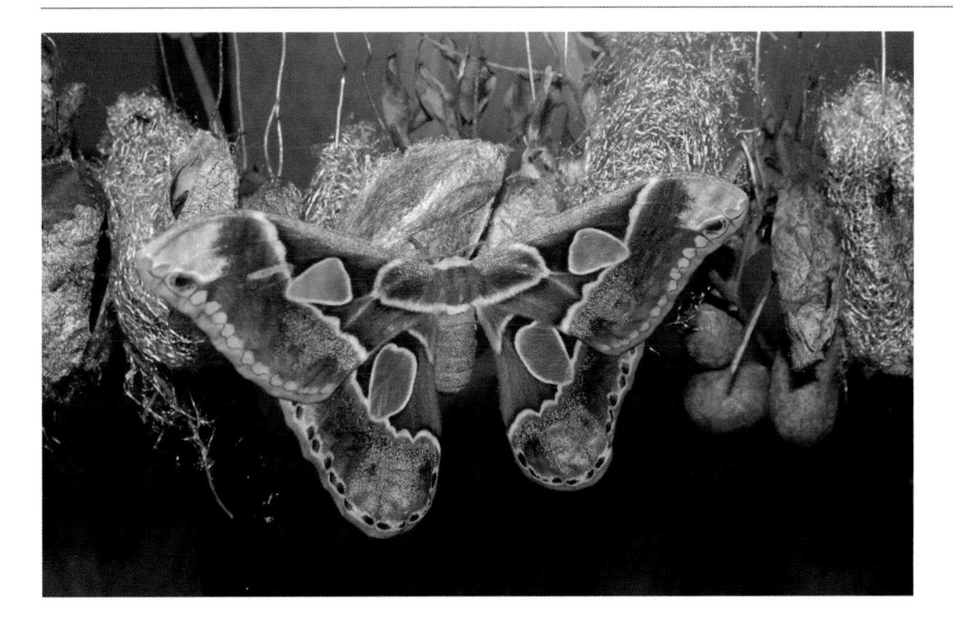

**Figure 7.3**
Ahimsa silk is cultivated in India. Ahimsa and peace silk allow the mulberry-fed moth to leave the cocoon before it is harvested.
*Udit Kulshrestha/Bloomberg via Getty Images*

- **Biodegradable:** All fibers, yarns, trims, and dyes used to manufacture the product or garment must also be biodegradable or disassembled before disposal. This should be substantiated with documentation that the product can completely break down into nontoxic material by being processed in a facility where compost is accepted. A secondary label or marketing material should be provided to instruct the customer.
- **Organic:** If organic silk is used.
- **Alternative dyes:** If used, such as natural dyes.
- **Hand wash in cold water or spot clean instead of dry-clean:** Can save on significant amounts of electrical energy over time.

# INNOVATION EXERCISES

## Design

1. Design a garment that emphasizes one or more of the sustainability benefits of silk.
2. Design a garment that circumvents one or more of the environmental impacts of silk.
3. Design a garment that encourages hand washing.
4. Develop an artisan project that would bring an additional social/Fair Trade element to a sustainable fabric program.
5. Create a garment using silk blended with another natural fiber that would allow you to emphasize the sustainability benefits of each fiber.
6. Develop a design concept that does not require ironing for wrinkles that would influence the customer to reduce ironing of the final product and the energy it uses.
7. Develop a design concept using lightweight silk in targeted areas that is supposed to degrade from exposure to sunlight and hot temperatures over time, encouraging the customer to engage with the garment and delay disposal.
8. Create a silk garment that is 100 percent biodegradable. Consider dyes and trims.

## Innovation Exercise Example

**Develop a design concept that does not require ironing for wrinkles that would influence the customer to reduce ironing of the final product and the energy it uses.**

**Figure 7.4**
In this example, smocking is used throughout to create a draped garment to hide the visible wrinkles.
*Amy Williams*

### Merchandising

9. Create a fifty-word marketing communication message for consumers about the sustainability benefits of silk.
10. Incorporating information from chapters 1, 2, 3, 4, 5, 6, and 7, develop a table that a retail buyer could use to quickly compare the advantages and disadvantages of 100 percent conventional cotton, 100 percent organic cotton, 100 percent flax, 100 percent bamboo linen, 100 percent hemp, 100 percent jute, 100 percent wool, and 100 percent silk.
11. Create a retail point-of-purchase sign or hangtag to promote fashion merchandise made from 100 percent silk.
12. Go online and find three fashion retailers who sell fashion merchandise made from 100 percent silk. What are the retailers and fashion merchandise offered? Why did they select this fiber for their fashion merchandise? How do they promote the advantages of this fiber to consumers? Do you believe their promotional strategies are accurate and effective? Why or why not?
13. Develop a marketing research protocol (e.g., focus group protocol, online survey, observational strategy) to determine consumers' perceptions of silk for fashion merchandise.

## REVIEW AND RECOMMENDATION

### Not Preferred

- Silk that is harvested on domesticated silk farms where the chrysalis is killed to prevent the moth from making a hole in the cocoon

### Preferred

- Wild or ahimsa silk, since the silkworm does not die in the process
- Silkworms that feed on organic mulberry leaves

### Recommended Improvements

- Incorporating environmental dyeing practices
- Developing end-of-use strategy (whether compostability or recyclability)

## OTHER SOURCES

1. Evripidis, Kipriotis. n.d. *The prospects of the European Sericulture within the frame of the EU Common Agricultural Policy.* Accessed March 28, 2015. www.bacsa-silk.org/en/the-prospects-of-the-european-sericulture-within-the-frame-of-the-eu-common-agricultural-policy/.
2. European Commission. n.d. *Agriculture and Rural Development.* Accessed March 28, 2015. ec.europa.eu/agriculture/quality/schemes/index_en.htm.
3. *WildSilkbase.* n.d. Accessed March 28, 2015. www.cdfd.org.in/wildsilkbase/info_moths.php.

Michael Norwood/iStock

# Leather

In the fashion industry, leather, suede, and nubuck from cows is most often used for clothing, shoes, and accessories (although leather can also be made from pigs, goats, sheep, and exotic animals). Many imitation leathers have emerged in the last decade for use in fashion (visit **Chapter 14: Imitation Leather** for more information), but most cannot duplicate leather's natural assets.

# Sustainable Benefits

- Due to the material's durability, leather garments can withstand decades of use and wear well, often looking even better with age. This physical and temporal durability can be seen to offset the sustainability impacts of a leather product, such as a jacket, handbag, or wallet, which is kept for a long time.
- Some leather suppliers only use leather that is a certified by-product of the meat industry (as in the case of cow, sheep, calf, and pig) (but not all leather used in the fashion industry is a by-product).

# Potential Impacts

## Chemical Use in Processing

The process by which animal skins are rendered soft, pliable, and durable enough for clothing and accessories is called **tanning**.

The chemicals used in the tanning process can be grouped into three categories: high, moderate, and low potential hazard. If these chemicals are not handled or disposed of properly, this could pose a serious risk to workers, their communities, and the environment.[1] Hazards can range from skin burns, eye injuries, and respiratory irritations to allergic reactions and fire hazards.

There are three stages whereby the skin of an animal can be formed into leather. The first stage is the preparation for tanning. The second stage is the actual tanning. The third stage applies finishing to the surface. Each stage poses potential social and environmental impacts.[2]

### Stage 1: Preparation Prior to Tanning

The skins are cured, dehaired, degreased, desalted, and soaked in an alkaline solution of sodium sulfide and lime. This makes the leather soft and flexible. The leather is then washed in running water and de-limed in a solution of ammonium chloride. A final bath of sulfuric acid renders a degree of acidity to the skin, which is helpful in the tanning process and also helps neutralize the alkali. This preparation process uses copious amounts of water and creates a high biological load in receiving water bodies if not treated.[3] It also causes **hydrogen sulfide** (a colorless, flammable, hazardous gas) emissions to air and water, which can be hazardous to workers.[4]

**BOX 8.1**
**SUSTAINABLE BENEFIT**

The physical and temporal durability of leather can be seen to offset the sustainability impacts of a leather product, if the product is kept for a long time.

**Figure 8.1**
Design leather items to be classic enough to transcend fashion trends.
*Michael Norwood/iStock*

To prevent bacterial growth damage to the animal skin during the soaking period, biocides, such as pentachlorophenol, are used. These substances have a high pollution index and are toxic to aquatic species.

### Stage 2: Tanning

Tanning is the process of converting putrescible skin into nonputrescible leather, usually with **tannin**, an acidic chemical compound that prevents decomposition and often imparts color. Tanning can be performed with either vegetable or mineral methods.

Mineral tanning usually uses chromium. Chrome tanning produces stretchable leather, which is excellent for use in handbags and garments. Chrome tanning includes water, salt, and chromic oxide and produces sulfur dioxide emissions to air. Sulfur dioxide is toxic to humans and other species. (Note: The type of chromium used in leather tanning is chromium(III) and should not be confused with the hexavalent chromium, chromium(IV). Chromium(IV) is a human carcinogen but does not have any tanning ability. Chromium(IV) may be found in chrome leathers in rare situations but can become more prevalent if the chrome-tanned leathers are bleached. Chrome tanning is completed in sodium and neutralized thoroughly in a solution of ammonium bicarbonate.

Chromium sulfate, zirconium, and aluminum mineral tanning substances may also be utilized to provide durability and strength to the raw skins.

Chrome tanning causes hydrogen sulfide emissions, polluting air and water if acidic effluent is not neutralized.[5]

Vegetable tanning uses tannin, which occurs naturally in vegetable matter and bark (chestnut, oak, tanoak, hemlock, quebracho, mangrove, wattle, and myrobalan). Vegetable-tanned hide is flexible and is used for some footwear, luggage, and furniture. Vegetable tanning is generally considered to have less impact on the environment, though rinsing and washing also use significant volumes of water, which must then be neutralized and treated for biological loads before being dispelled in the receiving water bodies. Table 8.1 shows the different tanning processes for leather.

**Table 8.1 Comparison Between Different Types of Tanning Processes for Leather**

| Type of Processing | Impacts | Detail |
|---|---|---|
| Chrome-tanned leather | • Tanned using chromium sulfate and other salts of chromium. | • Is more supple and pliable than vegetable-tanned leather and does not discolor or lose shape as drastically in water. A wider range and brighter colors are possible using chrome tanning. |
| Vegetable-tanned leather | • Tanned using tannin and other ingredients found in vegetable matter, tree bark, etc. | • Is stiffer than chrome leather and more porous to water. Colors are generally duller and more "transparent" than on chrome-tanned leathers. |
| Aldehyde-tanned leather (e.g., chamois leather) | • Tanned using **glutaraldehyde** or oxazolidine compounds. Uses emulsified oils, often those of animal brains. | • Referred to as wet-white leather due to its pale cream or white color. Exceptionally water absorbent and exceptionally soft. Often used for baby shoes because of its softness. Chamois leather is washable. |

### Stage 3: Finishing

After tannage, skiving or splitting of the skin to render it the correct thickness for commercial applications is a mostly mechanical process. The skin may in some cases be split into two useable pieces. In other cases, small pieces of leather are cut away from the final usable piece. These cut pieces can be processed into composite leather (visit **Chapter 14: Imitation Leather** for more information), which is generally stiff but does have limited commercial applications such as soles for footwear. This split leather is not considered 100 percent leather.

Finally, the hide may be waxed, rolled, lubricated, injected with oil, shaved, and dyed. Suedes and nubucks are finished by raising the nap of the leather by rolling with a rough surface.

Vegetable tanned leathers can also be given smooth, grained, bright, or dull finishes, all depending upon how the leather is hammered, folded, or rolled. Waterproofing also changes the final appearance of the leather.

## Consumer Care/Washing

Well-processed leather resists staining and can be wiped clean. Nubucks and suede require stain-resistant finishes, which are usually silicone sprays.

Some leather and suede garments are usually dry-cleaned, which can require the use of ozone-depleting chemicals.

- Visit **Appendix B: Consumer Care and Washing** for more guidance.

Chamois leather is washable, as long as the trims used in the garment are also selected to be washable.

## Reuse and Recyclability

During the production and patternmaking process, consider that leather can only be **downcycled** or reused for another product so do your best to eliminate pattern-cut waste.

- Visit **Chapter 26: Recycled/Circular Textiles Technologies** for more information about recyclability.

**Figure 8.2**
During the production and patternmaking process, consider that leather can only be downcycled, so do your best to eliminate pattern-cut waste.
*Denis Kartavenko/iStock*

**BOX 8.2 RECYCLABILITY**

Leather cannot be upcycled. While leather scraps can be bonded together with composites and spread out to make sheets, once the leather is bonded together, it is no longer recyclable and only fit for the landfill.

**Table 8.2 Optimize the Sustainability Benefits of Leather**

| Opportunity | Benefit | Considerations |
|---|---|---|
| Promote suppliers who have water treatment processes in place, including filtering for lime and chrome, catalytic sulfide oxidation, balancing, separation of particulate matter, appropriate sludge disposal, clarification of the effluent, and biological treatment. | • Ensures that water released to local waterways is clean. | • Finding out where your leather is sourced from might take some time and effort. Visit these suppliers to ensure that water treatment processes are in place, and if they are not, work with local nongovernmental organizations (NGOs) to set up. |
| Promote suppliers using low-chrome tanning. | • Chrome tanning is the most environmentally toxic part of the leather processing. Finding ways to minimize or circumvent this processing will improve sustainability impacts of leather. | • Finding out where your leather is sourced from might take some time and effort.<br>• Could result in different performance and durability qualities. Not as readily available as chrome-tanned leather. |
| Promote suppliers using nonchrome tanning processes such as wet-green®, which produces a tanning agent made from natural olive leaf extract. | • Chrome tanning is the most environmentally toxic part of the leather processing. Finding ways to minimize or circumvent this processing will improve sustainability impacts of leather. | • Leather is soft and durable and can be used for apparel, accessories, and furniture.<br>• Not as readily available as chrome-tanned leather and potentially more expensive. |
| Promote the use of organic leather. | • Typically, organic leather means that the hides are from animals that are organically fed and there is an environmental philosophy applied to the way the hide has been tanned (i.e., the hide has not been tanned with traditional chrome tanning). The water effluent is nontoxic. | • Not as readily available. Could produce a look/feel that is different from traditional chrome tanning.<br>• Will cost more than chrome-tanned leather. |
| Develop a policy to use leather from a certified by-product of the meat industry (as in the case of cow, sheep, calf, and pig). | • Ensures that fair animal practices are being maintained. | • Finding out where your leather is sourced from, and ensuring that the leather is a by-product of the meat industry, might take some time and effort. |

# Availability

Vegetable-tanned leather is readily available and applicable for accessory lines.

Low-chrome processes are becoming more available but tend to be more expensive than conventional chrome processing.

Many leather manufacturers, especially in Europe, are implementing state-of-the-art treatment processes and using this as a marketing advantage.

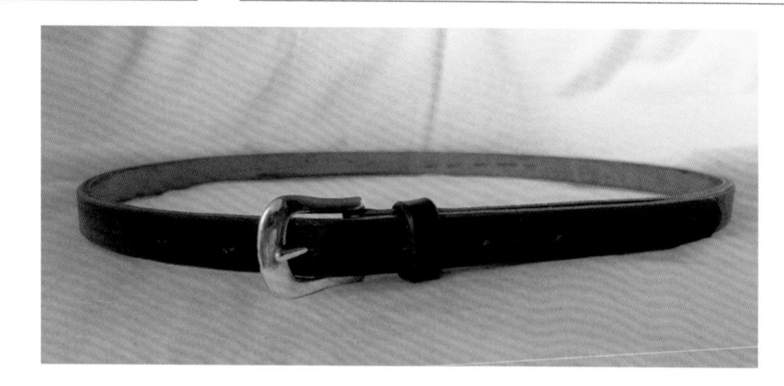

Organic leather is becoming available through the organic meat industry.

A tanning agent made from natural olive leaf extract is available from wet-green® in Germany.

**Figure 8.3**
Women's belt from the company Organic Leather. The leather comes from organically raised cows and is tanned with biodegradable tannins. www.organicleather.com.
*Organic Leather LLC*

**Figure 8.4**
Wet-green® is a nontoxic tanning agent that produces soft, supple leather that can be used for apparel, accessories, and furniture. www.wet-green.com/.
*www.wet-green.com/*

**BOX 8.3 ALTERNATIVE TO TRADITIONAL CHROME TANNING**

Wet-green® tanning agent is an alternative to traditional chrome tanning and is made from natural olive leaf extract. Wet-green® is Cradle to Cradle Certified GOLD-Level, a certification that verifies all chemicals used to make the agent have been assessed and verified safe.

## Fashion Applications

- Chrome tanning still provides the softest quality leather most suitable for clothing.
- Vegetable-tanned leather is best used for bags, belts, and some shoes.
- Aldehyde-tanned leather is applicable for baby shoes as well as adult clothing.
- Recycled composite leather is most applicable for belts and shoe soles, though it could be creatively applied to bags.

## Potential Marketing Opportunities

- **Vegetable tanned leather or "naturally tanned" leather:** If vegetable or naturally tanned.
- **Organic leather:** Organic leather is from certified organic meat. This can be claimed if used and verified.
- **Low-chrome tanning:** If low-chrome tanned.

# INNOVATION EXERCISES

### Design

1. Design a garment that emphasizes one or more of the sustainability benefits of leather.
2. Design a garment that circumvents one or more of the environmental impacts of leather.
3. Combine leather with other fabrics in the construction of the garment in order to minimize cost to the company and environment.
4. Create a garment or accessory where you strategically place leather at places of high stress on the garment to maximize its durable quality.
5. Design a leather item that is classic enough to transcend fashion trends.
6. Design a garment or accessory where you use recycled leather collected from tanneries to create modular accessories or patchwork pieces, or use in trims on garments.

### Merchandising

7. Create a fifty-word marketing communication message for consumers about the sustainability benefits of leather.
8. Incorporating information from Chapters 8 and 14, develop a table that a retail buyer could use to quickly compare the advantages and disadvantages of leather and imitation leather.
9. Create a retail point-of-purchase sign or hangtag to promote fashion merchandise made from leather.
10. Go online and find three fashion retailers who sell fashion merchandise made from leather. What are the retailers and fashion merchandise offered? Why did they select this material for their fashion merchandise? How do they promote the advantages of this material to consumers? Do you believe their promotional strategies are accurate and effective? Why or why not?
11. Develop a marketing research protocol (e.g., focus group protocol, online survey, observational strategy) to determine consumers' perceptions of leather for fashion merchandise.

## Innovation Exercise Example

**Use recycled leather collected from tanneries to create modular accessories or patchwork pieces, or use in trims on garments.**

**Figure 8.5**
In this example, long leather scraps are sewn together to create fringe at the bottom of a leather skirt.
*Amy Williams*

# REVIEW AND RECOMMENDATION

### Not Preferred

- Leather that has been traditionally chrome tanned

### Preferred

- Vegetable- or no-chrome-tanned leather

### Recommended Improvements

As you have learned in this chapter, the processing of leather can have pretty serious environmental and social impacts. In order to holistically capitalize on leather's best asset, its durability, be sure to maintain these practices:

- Incorporate closed-loop water effluent treatment processes to ensure processing chemicals have not been released untreated to local water bodies and drinking water.
- Design products using leather that are classic and intended to transcend changing trends.
- Use organic leather or leather that's a by-product of the meat industry.
- Ensure there is an end-of-use strategy.

# OTHER SOURCES

1. Leather Working Group: www.leatherworkinggroup.com/

acceleratorhams/iStock

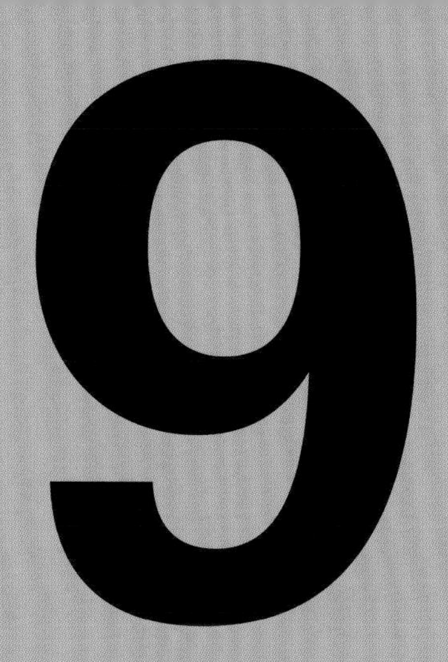

# Alpaca

The textile fiber obtained from alpaca is simply called "alpaca fiber." There are two different breeds of alpacas: huacaya and suri. The main physical difference between the two is the fleece. Like wool from sheep, fiber from huacaya has a natural crimp, and when the fleece grows out the animal looks fluffy. Suri fiber has no crimp in its fleece, and the fiber drapes down from the animal's body. It is soft and silky and can be used as an alternative to silk.[1]

# Sustainable Benefits

Unlike sheep or Kashmir goats, alpacas have padded feet and toenails rather than hooves, which means they are gentle on the land and do not contribute to land degradation and desertification like goats and sheep. Also, alpacas don't consume the root of grass like sheep and Kashmir goats, so the grass can continue growing after they have eaten it.[2]

Alpaca fiber is naturally warm and **water repellent**.

Alpaca fiber is highly valued for its softness, durability, and silkiness. Due to its low micron count (20–70 microns), it is very comfortable to wear and is also lightweight. It is naturally nonpilling.

The surface of alpaca fiber is water repellent, dirt repellent, and **stain repellent**, while the fiber interior is highly moisture absorbent, making it a comfortable fabric to wear.

Alpaca fiber absorbs odors and is, to some extent, self-cleaning. It tends not to smell bad, even after long periods of wear. Because of this, alpaca fiber garments and products do not need to be washed frequently.

Alpaca fiber comes in twenty-two natural colors, including whites, browns, grays, and black, potentially eliminating the need to use dyes.[3]

Unlike wool from sheep, alpaca fiber does not contain lanolin or grease, so it can easily be cleaned in a rinse bath with natural products. See

**BOX 9.1 SUSTAINABLE BENEFIT**

There are two different breeds of alpacas: huacaya (Figure 9.1) and suri (Figure 9.2). Huacaya fiber is fluffy like sheep's wool and can be used as an alternative to cashmere or wool, whereas suri fiber is silky and can be used as an alternative to silk.

**Figure 9.1**
Huacaya alpaca from the ranch Ahh . . . Sweet Alpacas in Vacaville, California.
*Annie Gullingsrud*

**Figure 9.2**
Suri alpaca from the ranch Ahh . . . Sweet Alpacas in Vacaville, California.
*Annie Gullingsrud*

**Chapter 6: Wool** for the impacts from scouring, a cleaning process to remove the lanolin grease from the wool fiber.

In 100 percent form, alpaca fiber fabric is biodegradable and can be broken down into simpler substances by microorganisms, light, air, or water in a nontoxic process. Absolute biodegradability depends on the dyes and trims used, as well as route of disposal.

- ♦ Visit **Appendix: Biodegradability in the Fashion Fibers STUDIO** for information about designing for biodegradability.

# Potential Impacts

## Consumer Care/Washing

Alpaca fiber fabrics may be hand washed, spot cleaned, or dry-cleaned, depending on the product.

- ♦ Visit **Appendix B: Consumer Care and Washing** for information for impacts related to consumer care and washing.
- ♦ Visit **Chapter 23: Dyeing and Printing** for more information on impacts associated with dyeing.

**Table 9.1 Optimize the Sustainability Benefits of Alpaca Fiber**

| Opportunity | Benefit | Considerations |
|---|---|---|
| Promote the use of alpaca fiber as an alternative to wool, cashmere, silk, and synthetics. | • Unlike sheep or Kashmir goats, alpacas have padded feet and toenails rather than hooves. Because of this, they are very gentle on the pasture and do not cause **desertification** like sheep or Kashmir goats.<br>• Alpacas don't consume the root of grass like sheep and Kashmir goats, so the grass can continue growing after they have eaten it.[4] | • Alpaca fiber is not as readily available as merino wool and will likely cost more. |
| Promote the use of alpaca fiber in children's clothes as an alternative to fabrics that need flame retardants. | • Alpaca fiber is a natural flame retardant, and it may be used as a viable alternative to synthetic fabrics that use fire retardants.[5]<br>♦ Visit **Chapter 24: Finishing** for more information. | • Alpaca fiber is not as readily available as synthetic fibers and costs more. |
| Develop relationships with producers and monitor farmers to ensure animal-friendly practices are being implemented. | • Making annual visits to the farms where you buy fiber will allow you to see firsthand that the animals are being treated properly.<br>• If you do find that best practices are not being maintained, in-person visits and good relationships with farmers will allow you to work with them to improve practices (because boycotting should be your last option!). | • Your company might purchase either yarn or the fabric, so it will be difficult to find out where the fiber has been sourced from (difficult, but not impossible).<br>• Visits require travel budgets. Build these visits into your annual budget. |
| Promote the use of natural color alpaca fiber. | • No bleaches or dyes are used in this case, and associated pollution impacts are avoided.<br>♦ Visit **Chapters 22 and 23** for more information about the impacts of bleaching and dyeing. | • Alpaca fiber comes in twenty-two natural colors, including white, browns, grays, and black. |

- Visit **Chapter 26: Recycled/Circular Textiles Technologies** for more information about recyclability.
- Visit **Appendix: Biodegradability in the Fashion Fibers STUDIO** for information about designing for biodegradability.

# Availability

Alpaca fiber is available from producers in Peru, North America, and Australia.

# Fashion Applications

Knitwear applications in clothing, accessories, outerwear, baby clothing, blankets, rugs, and upholstery.

**BOX 9.2**
**OPTIMIZE SUSTAINABILITY BENEFITS**

Promote the use of alpaca fiber in children's clothes as an alternative to fabrics that need flame retardants.[6]

**Figure 9.3**
Alpaca children's sweater.
*Stephen Shankland/iStock*

## Potential Marketing Opportunities

- **Renewable resource:** Alpaca fiber is a renewable natural resource and can be replenished by natural processes.
- **Biodegradable:** Undyed alpaca fiber is biodegradable. If claiming a garment is biodegradable, all fibers, yarns, trims, and dyes used to manufacture the product or garment must also be biodegradable or disassembled before disposal. This should be substantiated with documentation that the product can completely break down into nontoxic material by being processed in a facility where compost is accepted. A secondary label or marketing material should be provided to instruct the customer.
- **Undyed:** Promote the use of natural undyed alpaca fiber and the positive environmental impact it has due to the fiber not being dyed.

# INNOVATION EXERCISES

### Design

1. Design a garment that emphasizes one or more of the sustainability benefits of alpaca fiber.
2. Design a garment that circumvents one or more of the environmental impacts of alpaca fibers.
3. Design a garment using alpaca in blends with a less expensive natural fiber (wool, cotton, etc.) to help to minimize cost of the overall garment.
4. Design a garment using the natural shades of alpaca, including white, browns, grays, and black to create heathers with white wool or cotton.
5. Design a collection of at least five looks that use only undyed natural colors of alpaca fiber. What would be the sustainable benefit to using the fiber this way?
6. Design a garment that uses alpaca fiber in a strategic area to emphasize its self-cleaning and moisture-absorbent attributes.
7. Go to your local fabric or yarn store and find yarn or fabric made from 100 percent alpaca fiber. Identify the source of the fiber, including the name of the ranch and the location. Plan a visit if you live close!

### Merchandising

8. Create a fifty-word marketing communication message for consumers about the sustainability benefits of alpaca.
9. Incorporating information from chapters 1, 3, 6, 7, and 9, develop a table that a retail buyer could use to quickly compare the advantages and disadvantages of 100 percent conventional cotton, 100 percent organic cotton, 100 percent bamboo linen, 100 percent wool, 100 percent silk, and 100 percent alpaca.
10. Create a retail point-of-purchase sign or hangtag to promote fashion merchandise made from 100 percent alpaca.

## Innovation Exercise Example

**Design a garment that emphasizes one or more of alpaca's sustainability benefits. Alpaca fiber is highly valued for its softness, durability, and silkiness, making it perfect for children's sleepwear. Alpaca fiber is also naturally fire resistant, so topical chemical fire retardant coatings are not needed.**

- Visit **Chapter 24: Finishing** for more information on the impacts of fire retardants.

**Figure 9.4**
The design in this example shows a girl's 100 percent alpaca nightgown with alpaca trim, and matching alpaca booties.
*Amy Williams*

11. Go online and find three fashion retailers who sell fashion merchandise made from 100 percent alpaca. What are the retailers and fashion merchandise offered? Why did they select this fiber for their fashion merchandise? How do they promote the advantages of this fiber to consumers? Do you believe their promotional strategies are accurate and effective? Why or why not?
12. Develop a marketing research protocol (e.g., focus group protocol, online survey, observational strategy) to determine consumers' perceptions of alpaca for fashion merchandise.

# REVIEW AND RECOMMENDATION

## Preferred Fiber

Alpaca fiber is preferred for use as a sustainable fiber because:

- Alpaca is a renewable natural resource.
- The alpaca animals are gentle on the land and do not contribute to land degradation or desertification like sheep or Kashmir goats.
- Unlike wool from sheep, alpaca fiber does not contain lanolin or grease, so it can easily be cleaned in a rinse bath with natural products.
- It is naturally nonpilling and fire resistant and doesn't require nonpilling or fire-retardant chemical coatings.
- Alpaca fiber comes in twenty-two natural colors, including whites, browns, grays, and black, potentially eliminating the need to use dyes.
- It is biodegradable.

## Recommended Improvements

- Incorporating environmental dyeing practices
- Developing end-of-use strategy (whether compostability or recyclability)

## Can Be Used as an Alternative

The alpaca fiber can be used as an alternative to wool, cashmere, or even synthetics as a solution to some of the environmental challenges that these other fibers present.

svet110/iStock

# FUTURE FIBERS: NATURAL FIBERS

This Future Fibers part closer presents innovative biological alternatives to traditional biological fibers. From new processing techniques or new uses to biological fibers, the companies profiled here have created alternative-use fibers that can cut down on water use, chemical use, and water pollution.

## Pinatex™ [1]

Piñatex™ is a biological fiber derived from pineapple leaves. It is a nonwoven textile that is suitable as an alternative to leather.

### Sustainable Benefits

- Piñatex™ is breathable and soft, light and flexible, moldable, and easily dyed. It can be printed on and laser cut.

**Figure P1.1**
Piñatex™ is a biological fiber derived from pineapple leaves and is a suitable alternative to leather.
*Veg & City Drinks Ltd.*

- The by-product of decortication is biomass, which can be further converted into organic fertilizer or biogas. This can bring additional income to the pineapple farming communities.
- The fibers then undergo an industrial process to become a nonwoven textile, which is the base of Piñatex™.
- Piñatex™ is a by-product of the pineapple harvest, thus no extra water, fertilizers, or pesticides are required to produce Piñatex™ pineapple fibers.

### Potential Impacts

- At this time, is it unknown whether the chemicals used for dyeing the Piñatex™ fabric are safe, or what potential impacts the processing may incur.
- The Piñatex™ supply chain needs to be scaled up.

### Availability

The first samples of Piñatex™ will be available in the market in late 2016.

### Fashion Applications

This fabric may be used for fashion accessories, including shoes and handbags, upholstery, interiors, and the automotive industry.

## CRAiLAR[2]

**CRAiLAR** is a flax fiber that provides an alternative to cotton. After the flax growth cycle (refer to **Chapter 2: Flax** for more information about the environmental benefits of flax), the raw flax fiber is set to soften in the field. The raw fiber is then treated with the proprietary CRAiLAR process to remove

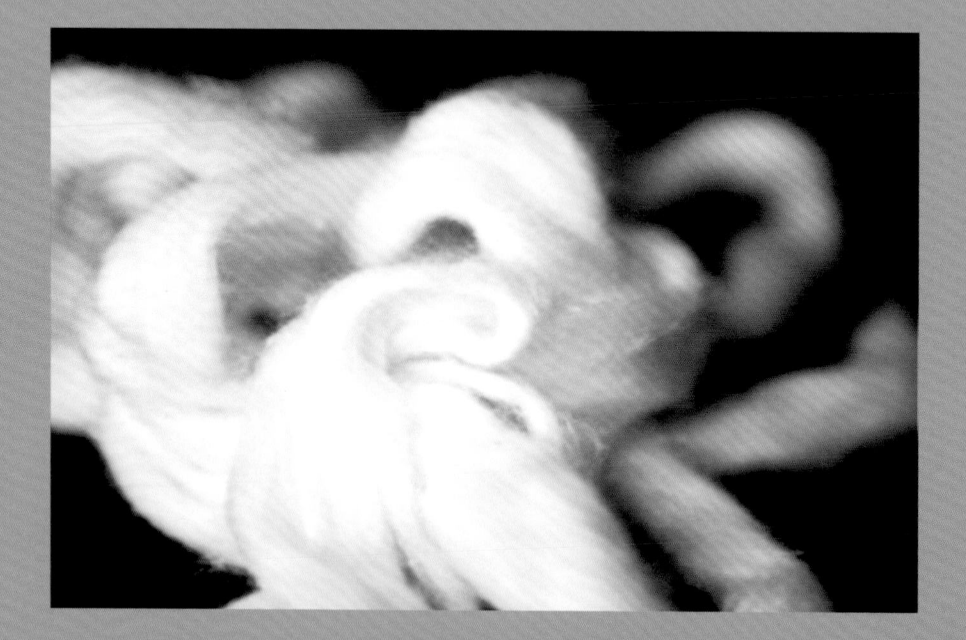

**Figure P1.2**
CRAiLAR flax fiber, after it has been treated with the company's proprietary process.
*Crailar Technologies, Inc.*

stiffening lignans from the flax, resulting in a fiber with cotton's softness and durability. Flax fibers treated with the CRAiLAR process are not chemically processed like rayon or viscose.

### Sustainable Benefits

- Flax is a good rotation crop, grows quickly, and requires few chemical inputs in its cultivation. See Part 1 Table 1 for a chart of fast-growing bast fibers.
- Some sources state that flax production requires half the volume of pesticides per acre compared to that of conventional cotton.[3]
- Flax is a rain-fed crop and generally doesn't require irrigation, where water has to be supplied to the land by means of ditches, pipes, or streams.[4]
- Flax may be grown organically without the use of toxic pesticides, and when claimed "organic" must meet the standard certification requirements by an internationally recognized certification agency accredited by the International Federation of Agriculture Movements (IFOAM).
- The innovation of the CRAiLAR enzymatic process application makes the CRAiLAR fiber soft and supple with a hand feel similar to cotton.

### Potential Impacts

#### Chemical Use in Cultivation

Flax does require herbicides to control weeds and, as a cellulosic fiber, it also requires some fertilizers. Synthetic fertilizers contain nitrogen salts that salinate the soil and over the long term decrease the productivity of the soil and may pollute aquatic ecosystems.

### Availability

CRAiLAR is currently available for large-scale production and is located in Canada.

### Fashion Applications

The softened CRAiLAR fiber is suitable for knitted applications whereas traditional flax fiber is coarse and has minimum flexibility, only allowing for woven textile production.

## Paper No. 9™

Paper No. 9™ is made from recycled paper fused with glue to "eco fabric." Organic cotton and remnants of wool, silk, and denim are then used as foundational backings to the paper. Paper No. 9™ stock collections are typically available with a denim or cotton twill backing, and the weight varies from 7.5 ounces to 16 ounces.

The backing changes depending upon the style and the desired color, texture, weight, and stiffness effect.

### Sustainable Benefits

- Provides "vegan" alternative to leather.
- Provides alternative to leather that has been chrome tanned.

**Figure P1.3**
Paper No. 9™ is a material made from recycled paper and glued to fabric. It provides an alternative to leather.
*Paper No. 9, LLC.*

- Recycled materials have been used.
- Attention has been put into the types of glues used for the process.
- These recycled papers are collected from local paper mills.
- There is no additional dyeing to process the papers.

## Potential Impacts

- Recyclability and compostability are still undetermined.
- The impact of glues and other adhesives is unknown and will need further assessment.

## Availability

Currently, Paper No. 9™ production is limited to small, one-of-a-kind pieces for runway. It is available through Paper No. 9 Company in Brooklyn, New York.

## Fashion Applications

Paper No. 9™ may be used for jackets, pants, skirts, blouses, trims, accessories, wallets, bags, light fixtures, wall coverings, and potentially shoes.

# Innovation Exercises

## Design

1. Design a garment that emphasizes one or more of the sustainability benefits of a biological future fiber.
2. Design a garment that circumvents one or more of the environmental impacts of biological future fiber.
3. Design a 100 percent biodegradable garment made from a biological future fiber.
4. Design a hangtag or point-of-sale (POS) material describing to the customer how to dispose of the item.
5. Develop a collection that is entirely designed to be easily deconstructed to enable a take-back and recycling program. Experiment with seaming and a variety of disassembly mechanisms in different fabrics.
6. Design an accessory or garment using a future biological fiber as an alternative to a traditional fashion fiber. Explain why using the future biological fiber circumvents negative impacts.

## Merchandising

7. Create a fifty-word marketing communication message for consumers about the sustainability benefits of Piñatex™ or CRAiLAR.
8. Incorporating information from Chapters 1 and 2 and Part 1 Future Fibers, develop a table that a retail buyer could use to quickly compare the advantages and disadvantages of 100 percent cotton, 100 percent organic cotton, 100 percent flax, 100 percent Piñatex™, and 100 percent CRAiLAR.
9. Create a retail point-of-purchase sign or hangtag to promote fashion merchandise made from a biological fiber (Piñatex™ or CRAiLAR).
10. Go online and find three fashion retailers who sell fashion merchandise made from Piñatex™ or CRAiLAR. What are the retailers and fashion merchandise offered? Why did they select this fiber for their fashion merchandise? How do they promote the advantages of this fiber to consumers? Do you believe their promotional strategies are accurate and effective? Why or why not?
11. Develop a marketing research protocol (e.g., focus group protocol, online survey, observational strategy) to determine consumers' perceptions of a biological fiber (Piñatex™ or CRAiLAR) for fashion merchandise.

*Aquafil S.p.A.-P.IVA*

## Part 2

# MANUFACTURED FIBERS

## Overview

Manufactured fibers are divided into three main classifications: cellulosic, protein (azlon), and synthetic fibers.

- **Manufactured synthetic fibers** are created using a polymerization process combining many small molecules into a large molecule (a polymer). Many of the polymers that constitute these fibers are similar to compounds that make up plastics, rubbers, adhesives, and surface coatings. Examples of synthetic fibers include polyester, nylon, spandex, imitation leather, acrylic, **polyethylene** and polypropylene.
- **Manufactured cellulosic fibers** account for approximately 8 percent of global man-made fibers. These fibers are derived from a range of plant-based and woody materials, which require intensive chemical manufacturing processes to be transformed first into pulp and then into "regenerated" cellulosic filaments. These fibers include modal, TENCEL® **Lyocell**, rayon/viscose made from bamboo, and rayon/viscose made from wood.
- **Manufactured protein fiber**, otherwise known as azlon, is composed of regenerated, naturally occurring protein derived from a number of sources, including soybean, peanut, casein (from milk), zein (from maize), and collagen/gelatin (from animal protein), to name a few. Protein fibers have received considerable attention in the United States, Europe, China, and Japan as an inexpensive substitute for wool, silk, and cashmere fibers.

The chapters in *Part 2: Manufactured Fibers* detail fibers in these categories that are prevalently used in the fashion industry. While some of these fibers can get a "bad rap," most of these fibers inherently possess sustainability attributes (recyclability being the main one). In most cases, all it takes is a little creativity and smart thinking to leverage these features. These chapters will guide you to make better decisions to optimize and identify potential innovation opportunities.

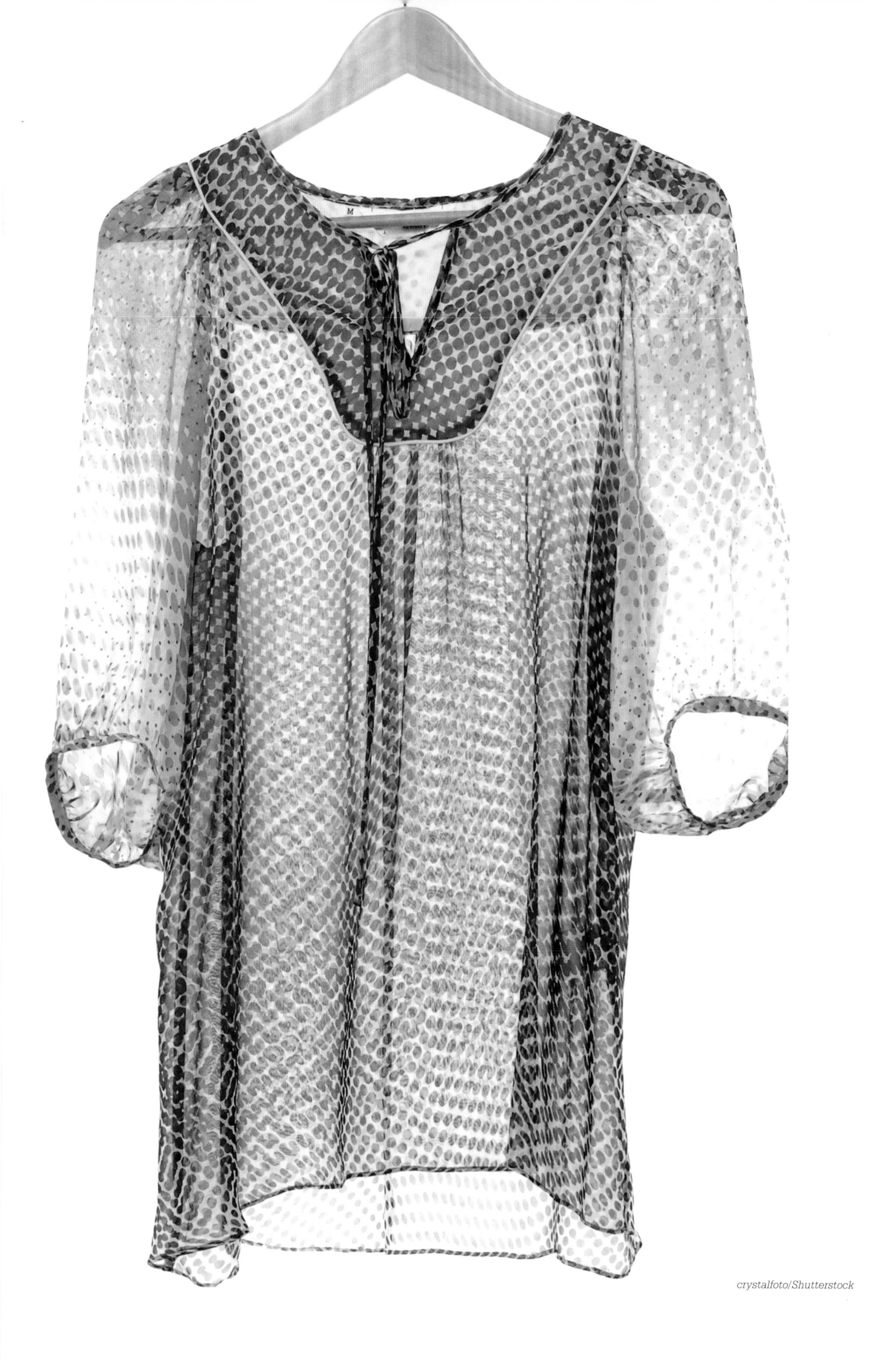

crystalfoto/Shutterstock

# 10

# Polyester

Over the last forty-five years, technical developments in polyester production have improved the fiber's hand feel, fineness, and quality. Polyester is now the world's most widely used fiber, representing 79 percent (in 2009) of world synthetic fiber production of synthetic fibers, fueled in part by its use in **fast-fashion** garments, the fastest growing sector of the fashion industry.[1,2,3] Europe's share in the polyester industry accounted for 960,000 tons in 2009–2010.[4]

Finding innovations that mitigate the ecological impacts of polyester will not only reduce environmental impacts but also potentially influence the textile industry as a whole.

Polyester is a manufactured, synthetic fiber. To produce polyester, crude oil (petroleum) is broken down into petrochemicals, which are then converted with heat and catalysts such as antimony into polyethylene terephthalate (PET). This is the same type of plastic used in plastic soda bottles.

# Benefits

Polyester fabrics are readily available, strong, resistant to stretching and shrinking, resistant to most chemicals, and don't easily succumb to wrinkling, mildew, or abrasion. When polyester fabrics are used in robustly constructed garments, they have the potential to last and to be worn many times, optimizing the embodied energy and resources in the garment. Refer to *Potential Impacts* that follows for a counterpoint to this benefit.

Polyester's positive attributes for clothing lie mostly in the consumer use phase of its life cycle, which accounts for 50–80 percent of a polyester garment's total ecological footprint. Polyester garments are generally washed in cold water and drip-dried, thereby minimizing water and energy use associated with garment care.[5]

In comparison to other synthetic fibers, there is currently more research and innovation when it comes to sustainability and improving polyester's environmental impact.

# Potential Impacts

## Processing

Petroleum, the main ingredient in manufacturing polyester, is a nonrenewable resource that takes millions of years to form and is currently being extracted from the earth for industrial uses faster than it can be replenished. The declining petroleum supply is the source of much debate—British Petroleum (BP) reports that there are 1,333 billion barrels still available to pump (enough for forty years at current usage rates).[6] Other sources state that supply is overestimated and that reserves are about 30 percent lower than widely reported.[7]

**Figure 10.1**
Spools of polyester at a factory in Bursa, Turkey.
*Salvator Barki/Gallo Images/Getty Images*

The manufacturing process for polyester is fully chemical, is energy intensive, and releases greenhouse gasses into the environment.[8]

In the production of polyester, the main ingredients used are terephthalic acid (TA) and dimethyl terephthalate, which are reacted with ethylene glycol, based on bromide-controlled oxidation.[9] The production of polyester releases emissions to air and water, which include heavy metal cobalt, manganese salts, sodium bromide, antimony oxide, and titanium dioxide.

Antimony is of particular concern, since it is a toxic heavy metal known to cause cancer under certain circumstances and is a suspected reproductive toxin.[10,11] The function of antimony in the production of polyester is as a catalyst in the oxidation process. But it is not necessary for polyester production, and alternate nonantimony catalysts are recommended but not yet readily available. There are preferred titanium-based catalysts being utilized in the market currently.

♦ Visit **Chapter 23: Dyeing and Printing** for more information on impacts associated with dyeing.

♦ Visit **Chapter 26: Recycled/Circular Textiles Technologies** for more information about recyclability.

♦ Visit **Appendix: Biodegradability in the Fashion Fibers STUDIO** for information about designing for biodegradability.

♦ Visit **Appendix B: Consumer Care and Washing** for more information.

# Optimize Sustainability Benefits

- Promote the use of polyester that has been recycled using a chemical process.
- Promote the use of mechanically recycled polyesters from producers that use high-quality raw materials.
- Promote the use of polyester catalyzed with a titanium catalyst.
- If using recycled polyester from PET bottles, ensure that supplier is using recycled bottles rather than new ones.
- Promote the use of low-impact dye and bleaching processes.

♦ Visit **Part 3: Processing** for information on bleaching and dyeing.

- Promote the use of OEKO-TEX® certified polyester.[12] OEKO-TEX® is an independent, third-party certifier that offers two certifications for textiles: OEKO-TEX® 100 (for products) and OEKO-TEX® 1000 (for production sites/factories). The OEKO-TEX® 100 label aims to ensure that products pose no risk to health. OEKO-TEX® certified products do not contain allergenic dyestuffs or dyestuffs that form carcinogenic aryl-amines. The certification process includes thorough testing for a long list of chemicals. Specifically banned are AZO dyes, carcinogenic and allergy-inducing dyes, pesticides, chlorinated phenols, extractable heavy metals, emissions of volatile components, and more.

# Availability

Due in part to the volume of discarded soda bottles, mechanically recycled polyester is readily available to textile and apparel suppliers. By far, the most common apparel product using this technology is polyester fleece, which several fabric suppliers now offer in 100 percent recycled polyester and in blends with virgin polyester and/or other fibers.

Companies such as Freudenberg Politex in Italy and REPREVE® and Poole Company in the United States are producing versions of mechanically recycled polyester that is of almost equal quality to virgin polyester because of the high quality of raw materials used.

**Chemically recycled** polyester is gaining in popularity, and the number of companies offering fabrics made from this technology is increasing globally. The Japanese company Teijin, which first developed chemical recycling technology, recently established a joint venture to establish fabric manufacturing in China.

Eco Intelligent™, **antimony-free** polyester, is available through Victor Group in North America.

**BOX 10.1**
**OPTIMIZE THE SUSTAINABILITY BENEFITS OF POLYESTER**

Promote the use of mechanically recycled polyesters from producers that use high-quality raw materials.

**Figure 10.2**
The North Face Denali jacket made with 100 percent recycled polyester using REPREVE® fibers.
*Clayto Boyd*

# Fashion Applications

- Chemically recycled polyester fibers maintain the same quality as virgin polyester fibers in perpetuity.
- Mechanically recycled polyester fibers can be of almost equal quality to virgin polyester, depending on the quality of raw materials. Some producers use low-quality materials, which result in low-quality fiber.
- Mechanically recycled polyester fibers can be blended with other fibers to maintain strength and quality for applications in a variety of fabric constructions—activewear, intimates, outdoor wear, T-shirts, trousers, etc.

# Potential Marketing Opportunities

- **100 percent recycled content:** Regulations require stating percent recycled if not 100% recycled content. In some cases where recycled polyester affects the aesthetic of the garment, craft marketing messages to turn potential negatives into positives.
- **Antimony free:** If nonantimony polyester is used.
- **Chemically recycled polyester:** If chemically recycled polyester is used. Make sure to explain the difference between chemically and mechanically recycled polyester on point-of-sale (POS) marketing, hangtags, and/or your website. Chemically recycled polyester supports closed-loop systems with materials that can be regenerated at virgin quality. This maximizes the embodied energy and resources in the garments.

# INNOVATION EXERCISES

### Design

1. Design a garment that emphasizes one or more of the sustainability benefits of polyester.
2. Design a garment that circumvents one or more of the environmental impacts of polyester.
3. Design a garment or product with reusable elements and for easy disassembly. Design the product so that trims, tags, buttons, etc., can be easily separated from the main body of the product at the end of its useful life, to enable easy in-house recycling.
4. Investigate alternative technologies for coloring polyester fabrics, such as AirDye, which eliminates water from the dyeing process. Design a garment in which you explore and feature unique aesthetics achieved from using this process.
5. Design a collection that is 100 percent polyester, including trims, so the garments can be chemically recycled easily at the end of their productive life.
6. Explore nondyeing mechanically recycled polyester to develop a unique aesthetic. Create a garment that features this aesthetic.

## Innovation Exercise Example

**Design a garment or product with reusable elements and for easy disassembly.**

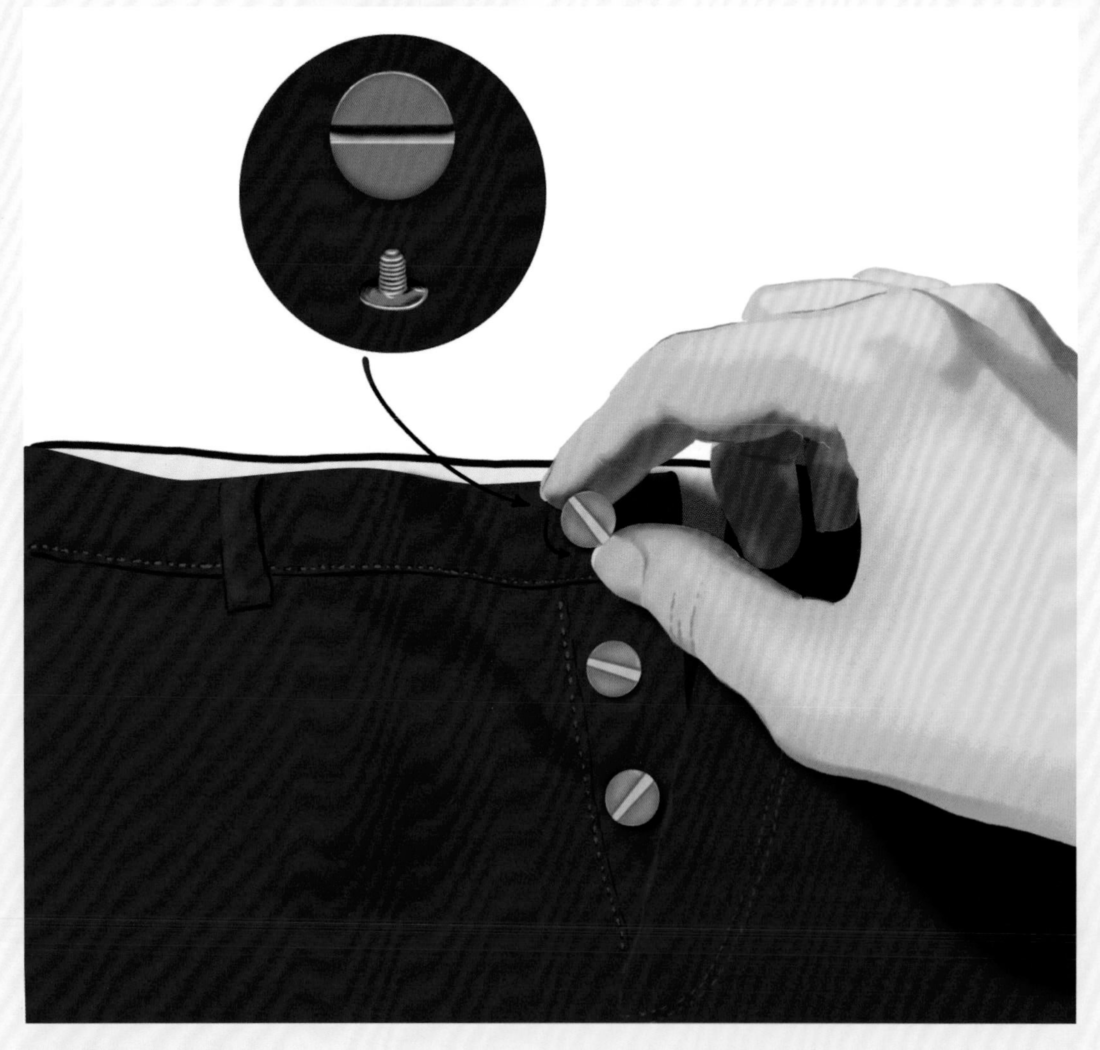

**Figure 10.3**
The design in this example shows how screw-in buttons allow metal buttons to be disassembled from polyester garments for reuse and recyclability of both materials.
*Yelena Safranova*

### Merchandising

7. Create a fifty-word marketing communication message for consumers about the sustainability benefits of recycled polyester.
8. Incorporating information from chapters 1, 3, 7, and 10, develop a table that a retail buyer could use to quickly compare the advantages and disadvantages of 100 percent conventional cotton, 100 percent organic cotton, 100 percent bamboo linen, 100 percent silk, 100 percent polyester, and 100 percent recycled polyester.
9. Create a retail point-of-purchase sign or hangtag to promote fashion merchandise made from 100 percent recycled polyester.
10. Go online and find three fashion retailers who sell fashion merchandise made from recycled polyester. What are the retailers and fashion merchandise offered? Why did they select this fiber for their fashion merchandise? How do they promote the advantages of this fiber to consumers? Do you believe their promotional strategies are accurate and effective? Why or why not?
11. Develop a marketing research protocol (e.g., focus group protocol, online survey, observational strategy) to determine consumers' perceptions of polyester and recycled polyester for fashion merchandise.
12. Develop a retail take-back and recycling program for polyester garments—including the objectives of the program and implementation strategies.

# REVIEW AND RECOMMENDATION

### Preferred

Polyester is a preferred sustainable fiber:

- Heavy-metal catalyst must be used.
- Polyester is recyclable, and if chemically recycled, can be upcycled into virgin-quality fiber.

### Recommended Improvements

- Use recycled polyester.
- Incorporate environmental dyeing practices.
- Develop end-of-use strategy to guarantee take-back and recyclability.

# OTHER SOURCES

1. Victor
2. Scaturro 2011
3. Braun and Levin 1986

gopfaster/iStock

11

# Nylon

There are several types of nylons, but the most widely utilized are nylon 6 and nylon 6,6.

Nylon 6 and 6,6 are manufactured, man-made fibers that are formed from a chemical process using carbon, hydrogen, oxygen, and nitrogen atoms. They differ in that they each begin with different polymer building blocks.[1]

The building blocks of nylon 6 and 6,6 are highly chemical and are derived from petroleum, a nonrenewable resource. Also, the fiber and its resulting fabric are nonbiodegradable. Efforts to address sustainability in these areas could help the overall influence of nylon on the environment.

# Benefits

Nylon 6 and 6,6 share a lot of the same fiber characteristics. They have strong wear resistance, abrasion resistance, chemical resistance, and heat resistance; are lustrous; have a high melting point; and are resilient.[2]

Nylon 6,6 has greater resilience, a higher melting point, and lower stain permeability than nylon 6, which makes nylon 6,6 perfect for carpet.[3]

The most notable characteristic of nylon 6 and 6,6 is their versatility. Although originally developed as an "artificial silk," both have been used for a vast variety of applications. Nylon fibers are used for garments, sheer hosiery, parachute cloth, backpackers' tents, bridal veils, musical strings, rope, broom and toothbrush bristles, Velcro . . . and many other applications![4]

Nylon 6 and 6,6 blend well with other fibers, and their chief contributions are strength and abrasion resistance.[5]

Nylon 6 and 6,6 are machine washable, dry quickly, need little pressing, and hold their shape well since they neither shrink nor stretch, thereby minimizing water and energy use associated with consumer care and washing.[6]

Due to their durability and abrasion resistance, some nylon 6 and 6,6 products have the potential to last and be worn many times, optimizing the energy and resources embodied in the product. For further details refer to the *Potential Impacts* discussion that follows.

# Potential Impacts

## Processing

Nylon 6 and 6,6 are made from petrochemical feedstock, which is a **nonrenewable resource**. Petroleum takes millions of years to form and is currently being extracted from the earth for industrial uses faster than it can be replenished.

**Durable water repellents (DWRs)** are applied to nylon 6 and 6,6 garments and products to allow for breathability and water repellency. Fluorochemicals are commonly used in these water-repellent finishes and waterproof membranes (thin films or coatings attached to the back of the fabrics to prevent water from passing through). Two fluorinated compounds, perfluorooctane sulfonate (PFOS) and perfluorooctanoic acid (PFOA), are of the most concern since they are known to have persistent bioaccumulative and toxicological effects on the environment. The European Union has banned PFOS, and some countries in the EU have also banned PFOA.[7]

Waterproof membranes are engineered to be breathable. They are commonly derived from petroleum and are often made using PFOA.

## Consumer Care and Washing

Nylon products are typically machine washed. Certain at-home detergents have been reported to have detrimental effects on humans and the environment; they contribute to ozone depletion and can pollute wastewater.

- Visit **Appendix B: Consumer Care and Washing** for more information.
- Visit **Chapter 23: Dyeing and Printing** for more information on impacts associated with dyeing.

- Visit **Chapter 26: Recycled/Circular Textiles Technologies** for more information about recyclability.
- Visit **Appendix: Biodegradability in the Fashion Fibers STUDIO** for information about designing for biodegradability.

# Biodegradability and Recyclability

## Not Biodegradable

Nylon 6 and 6,6 are synthetic fibers from a carbon-based chemical feedstock and are considered **nonbiodegradable**; they cannot decompose naturally by living organisms or bacteria.[8]

## Recyclable (Commercially)

Nylon is recyclable and is currently being recycled commercially.

Using recycled nylon achieves two main ecological benefits: 1) it slows the depletion of virgin natural resources, and 2) it reduces textile waste building in landfills. Nylon can be recycled into new versions of the same product or into entirely different products.

**Post-consumer waste** from used and discarded products and **post-industrial waste** from material collected during the product manufacturing can be recycled. There are two processes for recycling nylon fibers: mechanical and chemical.

### Mechanical Recycling

Nylon 6 and 6,6 can be effectively collected, cleaned, cut, remelted, and remolded to make yarns. However, the fiber is "downcycled" in this manner, which means that its physical structure breaks down, and eventually the product must be discarded to landfill.[9]

Collecting, sorting, and purifying discarded synthetic garments (i.e., post-consumer waste) is currently cumbersome. Infrastructures for labeling, collecting, and sorting need to be improved so that the post-consumer raw material source can scale to be economically viable.

### Chemical Recycling

There is potential for nylon 6 to be chemically recycled. Chemical recycling involves breaking the polymer into its molecular parts and reforming the molecules into a yarn of equal strength and quality as the original, in perpetuity.[10] In this process, the chemical building blocks are separated (depolymerization) and reassembled (repolymerization), forming what is known as a "**closed loop**" where the final stage of the product's life cycle (disposal) forms the first stage of the next product (raw fiber). Closed-loop recycled nylon processing is currently limited to nylon 6. In addition, the infrastructure to label, collect, sort, and purify discarded garments must be in place.[11]

- Visit **Chapter 26: Recycled/Circular Textiles Technologies** for more information about recyclability.

# Optimize Sustainability Benefits

- Promote the use of chemically recycled, closed-loop nylon 6.
- Support developments of chemically recycled nylon 6,6.
- Investigate developments in bio-nylon 6, which uses amino acids derived from dextrose fermentation as the starting material, instead of petroleum.
- Investigate nontoxic flame-retardant applications for nylon.
- Promote the use of halogen-free flame retardants.
- Investigate nontoxic waterproofing methods for nylon.
- Investigate alternative technologies for coloring nylon fabrics, such as AirDye, which eliminates water from the dyeing process.[12]
- Promote OEKO-TEX® certified nylon.[13] OEKO-TEX® is an independent, third-party certifier that offers two certifications for textiles: OEKO-TEX® 100 (for products) and OEKO-TEX® 1000 (for production sites/factories). OEKO-TEX® 100 labels aim to ensure that products pose no risk to health. OEKO-TEX® certified products do not contain allergenic dyestuffs or dyestuffs that form carcinogenic aryl-amines. The certification process includes thorough testing for a long list of chemicals. Specifically banned are AZO dyes, carcinogenic and allergy-inducing dyes, pesticides, chlorinated phenols, extractable heavy metals, emissions of volatile components, and more.
- Investigate PFOS- and PFOA-free water repellents.
- Investigate using waterproof membranes from renewable resources.

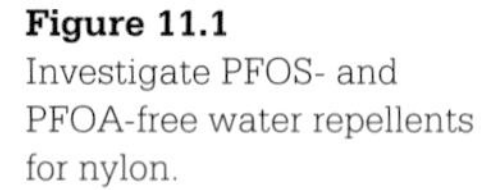

**Figure 11.1**
Investigate PFOS- and PFOA-free water repellents for nylon.
*D.R. Hutchinson/Getty Images*

# Availability

OEKO-TEX® Standard 100 certified nylon is available. Manufacturers can be found at www.oeko-tex.com.

Nontoxic methods of waterproofing and flame retardants are available.

Recycled nylon is available globally in the United States, Europe, Slovenia, Croatia, China, Japan, and Israel.[14,15] ECONYL® Recycled nylon is available from Aquafil in Italy.

# Fashion Applications

Nylon can be used for jackets, lingerie, swimwear, exercise wear, hosiery, bedspreads, carpets, upholstery, tents, fish nets, sleeping bags, rope, parachutes, and luggage.

Some companies are producing versions of mechanically recycled nylon that are of almost equal quality to virgin nylon.

# Potential Marketing Opportunities

- **X percent recycled content:** If using recycled content, state percentage used, if not 100 percent.
- **Nontoxic DWR methods:** If used and verified.

**Figure 11.2**
Close-up of nylon hosiery.
*Cultura RM Exclusive/Gregory S. Paulson*

**BOX 11.1 RECYCLED NYLON**

Italian company Aquafil has developed a recycled nylon, ECONYL®, from recycling nylon fishing nets. The company's next step will be to create recycled nylon from post-consumer waste.

**Figure 11.3**
Jacket from Outerknown made form ECONYL® Recycled nylon. www.outerknown.com
*Aquafil S.p.A.-P.IVA*

- **Nontoxic methods of waterproofing:** If used and verified.
- **OEKO-TEX® Standard 100 certified:** If used and verified.
- **X percent bio-based:** If used and verified.

# INNOVATION EXERCISES

## Design

1. Design a garment that emphasizes one or more of the sustainability benefits of nylon 6 or 6,6.
2. Design a garment that circumvents one or more of the environmental impacts of nylon 6 or 6,6.
3. Design a collection of garments that are 100 percent recyclable.
4. Create internal store collection to take back nylon 6 and 6,6 garments and products. Use fabric from collected garments and products to innovatively redesign new products and prolong their life cycle.
5. Design a garment from recycled nylon that features its unique aesthetic.

## Merchandising

6. Create a fifty-word marketing communication message for consumers about the sustainability benefits of nylon.

7. Incorporating information from chapters 1, 3, 7, 10, and 11, develop a table that a retail buyer could use to quickly compare the advantages and disadvantages of 100 percent conventional cotton, 100 percent organic cotton, 100 percent bamboo linen, 100 percent silk, 100 percent polyester, 100 percent recycled polyester, 100 percent nylon, and 100 percent recycled nylon.
8. Create a retail point-of-purchase sign or hangtag to promote fashion merchandise (including home fashions) made from 100 percent recycled nylon.
9. Go online and find three fashion retailers who sell fashion merchandise made from recycled nylon (including home fashions). What are the retailers and fashion merchandise offered? Why did they select this fiber for their fashion merchandise? How do they promote the advantages of this fiber to consumers? Do you believe their promotional strategies are accurate and effective? Why or why not?
10. Develop a marketing research protocol (e.g., focus group protocol, online survey, observational strategy) to determine consumers' perceptions of nylon and recycled nylon for fashion merchandise.
11. Develop a retail take-back and recycling program for nylon products—including the objectives of the program and implementation strategies.

# REVIEW AND RECOMMENDATION

### Preferred

Nylon is preferred because:

- Nylon 6 has a single monomer, caprolactam, which is no more than moderately toxic to humans or environment.
- Nylon 6 is recyclable.

### Recommended Improvements

- Incorporate nontoxic waterproofing methods for nylon.
- Use recycled nylon.
- Incorporate environmental dyeing practices.
- Develop end-of-use strategy to guarantee take-back and recyclability.

ericsphotography/iStock

# 12

# Spandex

Spandex—also known as elastane—is the generic name for the synthetic, manufactured fiber that is a long chain synthetic polymer. Some trade names for these fibers include LYCRA® (DuPont) and Dorlastan® (Bayer).[1]

When used as the central filament core yarn with staple fibers such as cotton, spandex becomes the silent hero in the consumer use stage, as it can prolong the use of a product by helping it retain its shape. The manufacturing process of spandex, however, is energy intensive and highly chemical, and the building blocks are derived from petroleum, a nonrenewable resource. Also, spandex is often used blended with other fibers, such as cotton. While a blend with spandex enhances performance qualities of other fibers, the spandex will prevent biodegradability when blended with natural fibers and also prevent recyclability when blended with other synthetic fibers. Efforts to address sustainability in these areas could help the overall influence of spandex on the environment.

# Benefits

Spandex was developed as an alternative to traditional natural fibers, since it can stretch and then snap back to its original form, whereas most natural fibers cannot.

It is lightweight, soft, and smooth, and it does not restrict movement. It can be easily dyed and is resistant to abrasion, body oils, and perspiration.

The entirety of spandex's environmental benefits occurs in the consumer use phase. Spandex can be stretched repeatedly—up to 500 percent of its length—and will return to its original size.[2]

What makes spandex special is its compatibility with other fibers or yarns. With just a small percentage in the overall fabric composition, spandex can prolong the life of a product by helping it retain its shape. Spandex fibers are blended in textile applications where high elastic extension and recovery are needed for the material produced—such as stretch denim, activewear, and underwear.[3] This allows for less stress on seams and helps prevent garments from becoming loose fitting in high stress areas such as the elbows or knees. This feature of spandex can help prolong the use of the product and divert waste from landfills.[4] This is discussed under the *Environmental Impact* section that follows.

Spandex can be machine washed and drip-dried, depending on the other fibers it is combined with, thereby minimizing water and energy use associated with consumer care and washing.

# Potential Impacts

## Processing

All spandex fibers are segmented polyurethane and formed through spinnerets either by melt extrusion or by solvent spinning.[5] Generally, polyurethane used for spandex is a by-product of petroleum, which is a nonrenewable resource.

**Figure 12.1**
Spanx represents the best of spandex, allowing it to stretch, shape to the body, and snap back to its original form, whereas most natural fibers cannot.
*Amy Sussman/Getty Images for Spanx*

**Figure 12.2**
Currently, spandex blended with either a natural or synthetic fiber is neither biodegradable nor recyclable.
*AlexTois/iStock*

Petroleum takes millions of years to form and is currently being extracted from the earth for industrial uses faster than it can be replenished. The declining petroleum supply is the source of much debate—British Petroleum (BP) reports that there are 1,333 billion barrels still available to pump (enough for forty years at current usage rates).[6] Other sources state that supply is overestimated and that reserves are about 30 percent lower than widely reported.[7]

Many common solvents used for spandex production are toxic. Solvents such as dimethylformamide (DMF) are potent liver toxins, and research points to a possible association with cancer.[8]

The production of spandex emits hazardous pollutants to air, which include toluene and 2,4-toluene diisocyanate (TDI).[9] Toluene is found in gasoline, acrylic paints, varnishes, lacquers, paint thinners, adhesives, glues, rubber cement, airplane glue, and shoe polish. Although they are not characterized as carcinogens, **chronic** inhalation exposure to toluene and 2,4-toluene diisocyanate in workers has caused significant decreases in lung function and an asthma-like reaction.[10] Toluene levels of 500 ppm are considered immediately dangerous to health and life.[11]

- Visit **Chapter 23: Dyeing and Printing** for more information on impacts associated with dyeing.
- Visit **Chapter 26: Recycled/Circular Textiles Technologies** for more information about recyclability.
- Visit **Appendix: Biodegradability in the Fashion Fibers STUDIO** for information about designing for biodegradability.

# Optimize Sustainability Benefits

- Use spandex that has been certified by a third-party, multi-attribute certification or standard, such as OEKO-TEX® certified spandex.[12] OEKO-TEX®

is an independent, third-party certifier that offers two certifications for textiles: OEKO-TEX® 100 (for products) and OEKO-TEX® 1000 (for production sites/factories). OEKO-TEX® 100 labels aim to ensure that products pose no risk to health. OEKO-TEX® certified products do not contain allergenic dyestuffs or dyestuffs that form carcinogenic aryl-amines. The certification process includes thorough testing for a long list of chemicals. Specifically banned are AZO dyes, carcinogenic and allergy-inducing dyes, pesticides, chlorinated phenols, extractable heavy metals, emissions of volatile components, and more.

- Use spandex derived from bio-based sources, rather than petroleum. Genomatica, a process technology developer for the chemical industry located in the United States, has developed a process that converts sugar (derived from sugarcane or beets or others) into commercial grade 1,4-butanediol (BDO), known as the GENO BDO™ process. BDO is a precursor to the chemical that makes spandex fibers. Bio-based BDO produced from the GENO BDO™ process is made from renewable feedstocks, rather than a conventional BDO made with petroleum-based feedstocks. Genomatica is currently licensing its process technology to producers and users in the chemical industry.[13] While this spandex is partially bio-based (deriving from natural materials), it is not biodegradable.
- INVISTA is also a bio-based spandex, and 70 percent of the weight of the fiber comes from a renewable source made from dextrose, derived from corn. While this spandex is partially bio-based (deriving from natural materials), it is not biodegradable.[14]

# Availability

- OEKO-TEX® Standard 100 certified spandex is available. Manufacturers can be found at https://www.oeko-tex.com/.
- Bio-based spandex is available by contacting Genomatica to be connected with suppliers that license the GENO BDO™ process.
- INVISTA is readily available.

# Fashion Applications

- Covered elastic yarn (covered with a spun or filament yarn to hide the spandex yarn): heavyweight foundations, elastic bandages, athletic supporters.[15]
- Bare elastic yarn (monofilament spandex fiber): swimwear, athletic wear, lightweight foundation garments.[16]
- Core spun yarns (central filament core with staple fiber): active sportswear, stretch denim, stretch chino.[17]

**Figure 12.3**
Spandex swimsuit.
*Randy Brooke/Wire Image/Getty Images*

# Potential Marketing Opportunities

- **OEKO-TEX® Standard 100 certified:** OEKO-TEX® certified fiber must be used and verified.
- **X percent bio-based:** Market this on hangtags by featuring the percentage of bio-based material used and type, or market the brand name of the fiber, such as INVISTA.

# INNOVATION EXERCISES

## Design

1. Design a garment that emphasizes one or more of the sustainability benefits of spandex.
2. Design a garment that circumvents one or more of the environmental impacts of spandex.
3. While recycling of spandex and spandex blended with other fibers is being developed, design an interim solution for recycling.
4. Design a garment or product with reusable elements and for easy disassembly. How will the spandex blended fabric be reused?

## Merchandising

5. Create a fifty-word marketing communication message for consumers about the sustainability benefits of bio-based spandex.
6. Incorporating information from chapters 1 and 12, develop a table that a retail buyer could use to quickly compare the advantages and disadvantages of 100 percent conventional cotton, 100 percent organic cotton, 95 percent conventional cotton/5 percent spandex, and 95 percent organic cotton/5 percent bio-based spandex.
7. Create a retail point-of-purchase sign or hangtag to promote fashion merchandise made from 95 percent organic cotton/5 percent bio-based spandex.
8. Go online and find three fashion retailers who sell fashion merchandise made from bio-based spandex blended with another fiber. What are the retailers and fashion merchandise offered? Why did they select this fiber blend for their fashion merchandise? How do they promote the advantages of this fiber blend to consumers? Do you believe their promotional strategies are accurate and effective? Why or why not?
9. Develop a marketing research protocol (e.g., focus group protocol, online survey, observational strategy) to determine consumers' perceptions of bio-based spandex blended with other fibers for fashion merchandise.

# REVIEW AND RECOMMENDATION

## Not Preferred

Spandex is not preferred for use as a sustainable fiber because:

- Spandex is neither recyclable nor biodegradable.
- Many common solvents used for spandex production are toxic. Solvents such as dimethylformamide are potent liver toxins, and research points to a possible association with cancer.
- The production of spandex emits hazardous pollutants to air, which include toluene and 2,4-toluene diisocyanate. Toluene is found in gasoline, acrylic paints, varnishes, lacquers, paint thinners, adhesives, glues, rubber cement, airplane glue, and shoe polish. Although they are not

characterized as carcinogens, chronic inhalation exposure to toluene and 2,4-toluene diisocyanate in workers has caused significant decreases in lung function and an asthma-like reaction. Toluene levels of 500 ppm are considered immediately dangerous to health and life.

## Recommended Improvements

- Encourage the development of new biodegradable elastomers that can be suitably combined with cotton for denim.
- Support optimization of chemical processes for spandex.

morrolight/iStock

13

# Acrylic

Acrylic is a manufactured synthetic fiber in which the fiber-forming substance is any long-chain synthetic polymer.[1] Acrylic production is a multistep chemical process that uses 85 percent acrylonitrile. Acrylonitrile is made through a chemical process from propylene gas.[2] Propylene is a gas that is produced from the by-product of refinery operations.[3]

Acrylic represents a large portion of overall synthetic fiber use, although recent reports have indicated that it is gradually decreasing in usage. In 2010, world production of acrylic fibers was at 3 million tons, most produced in China and Europe.[4] In 2012, 567,000 tons of acrylic fiber was produced in Europe.[5]

# Benefits

Acrylic is reasonably comfortable and has a wool-like aesthetic but is less expensive than wool.

Due to their durability and excellent sunlight and general resistance to wear and tear, acrylic yarn and fabrics have the potential to last and be worn many times, optimizing the energy and resources embodied in the garment. This is also an environmental impact (see *Potential Impacts* for further details).

Acrylic is machine washable, and acrylic products are generally washed in cold water and drip-dried. Unlike wool, acrylic won't shrink when washed, and it doesn't require antishrinking treatments that may be toxic.

- Visit **Chapter 6: Wool** for more guidance.

# Potential Impacts

### Processing and Chemical Treatments

Acrylic fibers contain at least 85 percent acrylonitrile. Studies done by the **United States Environmental Protection Agency (EPA)**, whose mission is to protect human health and the environment, show that workers repeatedly breathing small amounts of acrylonitrile over long periods of time may develop cancer. Acrylonitrile enters the body through inhalation or absorption through skin contact.[6] The Centers for Disease Control and Prevention (CDC) suggests preventing skin contact.[7] This is unsettling because often acrylic is worn against the skin!

Acrylic processing emits high amounts of **volatile organic compounds (VOCs)**, which are released as gases from certain solids or liquids. Air emissions from the acrylic process include volatilized residual monomer, organic solvents, additives, and other organic compounds used in fiber processing.[8]

### Chemical Treatments for Pilling

Acrylic fibers are highly likely to pill. This can cause wearers to donate or dispose of acrylic garments in a short amount of time. It's also worth pointing out that most people might not want to buy a pilled sweater at a secondhand store.

To address this natural pilling, acrylic fabrics are sometimes chemically treated during the manufacturing process in order to reduce their propensity to pill. If used, these chemicals need to be investigated to ensure safety, both for the worker and the wearer.

- Visit **Chapter 23: Dyeing and Printing** for more information on impacts associated with dyeing.
- Visit **Chapter 26: Recycled/Circular Textiles Technologies** for more information about recyclability.
- Visit **Appendix: Biodegradability in the Fashion Fibers STUDIO** for information about designing for biodegradability.

# Optimize Sustainability Benefits

- When using acrylic fiber, use it intentionally and understand that its impact on the environment may go far beyond its processing—acrylic is generally not being recycled and is not biodegradable.

- Use acrylic fiber that has been certified by a third-party, multi-attribute certification or standard, such as OEKO-TEX® certified acrylic.[9] OEKO-TEX® is an independent, third-party certifier that offers two certifications for textiles: OEKO-TEX® 100 (for products) and OEKO-TEX® 1000 (for production sites/factories). OEKO-TEX® 100 labels aim to ensure that products pose no risk to health. OEKO-TEX® certified products do not contain allergenic dyestuffs or dyestuffs that form carcinogenic aryl-amines. The certification process includes thorough testing for a long list of chemicals. Specifically banned are AZO dyes, carcinogenic and allergy-inducing dyes, pesticides, chlorinated phenols, extractable heavy metals, emissions of volatile components, and more.

**Figure 13.1**
Natural pilling in acrylic.
*Miyuki Satake/iStock*

# Availability

Most of the global acrylic fiber capacity is in the Asian region, with a concentration in China. North America and West Europe together now create less than 20 percent of the global volume of acrylic.[10] OEKO-TEX® certified acrylic is available in China.

**BOX 13.1**
**SUSTAINABLE BENEFIT**

Consider that pilling can actually be desirable! Create a design that emphasizes naturally pilling of fiber to circumvent chemical treatments and celebrate natural pilling of acrylic fiber and promote longevity of use and wear.

# Fashion Applications

Acrylic is commonly used in sweaters, women's and children's clothing, sportswear, socks, knitted underwear, pajamas, gloves, carpets, rugs, upholstery, cushions, blankets, outdoor umbrellas, and tents.[11]

# Potential Marketing Opportunities

- **OEKO-TEX® Standard 100 certified:** Verify that you are using OEKO-TEX® certified acrylic.

**Figure 13.2**
Use acrylic fiber that has been certified by a third-party, multi-attribute certification or standard, such as OEKO-TEX® certified acrylic.[12]
*Deepak Aggarwal/Dorling Kindersley/Getty Images*

# INNOVATION EXERCISES

## Design

1. Design a garment that emphasizes one or more of the sustainability benefits of acrylic.
2. Design a garment that circumvents one or more of the environmental impacts of acrylic.
3. What fabrics would you use as an alternative to acrylic? What sustainable decisions would you make around these alternative fabrics to ensure sustainable practices are carried through from cultivation to processing?
4. Design a garment and products with reusable elements—such as trims and tags. Design the product so that trims and tags can be easily separated from the main body of the product at the end of use, to enable easy recycling.
5. Currently, acrylic recycling is not commercially available. While this technology is being developed, think about how you could extend the use of acrylic garments by redesigning them into new products. Design a garment where you use fabric from collected garments and products to innovatively redesign a new product or garment.
6. Create garments where the natural pilling that occurs with wear is intentional and celebrated to increase use of the product and divert waste from landfills.

## Merchandising

7. Create a fifty-word marketing communication message for consumers about the sustainability benefits of acrylic.
8. Incorporating information from chapters 6 and 13, develop a table that a retail buyer could use to quickly compare the advantages and disadvantages of 100 percent wool and 100 percent acrylic.
9. Create a retail point-of-purchase sign or hangtag to promote fashion merchandise made from 100 percent OEKO-TEX® Standard 100 certified acrylic.
10. Go online and find three fashion retailers who sell fashion merchandise made from OEKO-TEX® Standard 100 certified acrylic. What are the retailers and fashion merchandise offered? Why did they select this fiber for their fashion merchandise? How do they promote the advantages of this fiber to consumers? Do you believe their promotional strategies are accurate and effective? Why or why not?
11. Develop a marketing research protocol (e.g., focus group protocol, online survey, observational strategy) to determine consumers' perceptions of acrylic and OEKO-TEX® Standard 100 certified acrylic for fashion merchandise.

# REVIEW AND RECOMMENDATION

## Not Preferred

Acrylic fiber is not preferred for use as a sustainable fiber. Currently, the environmental impacts of acrylic greatly outweigh its benefits. This is generally due to its environmental impacts during the processing stage and the fact that acrylic, although technically recyclable, is not readily being recycled at this time. This lack of balance indicates the necessity of exploring more sustainable options—from using other fibers entirely to working with partners to develop recycling capabilities for acrylic.

## Innovation Exercise Example

**Create garments where the natural pilling that occurs with wear is intentional and celebrated to increase use of the product and divert waste from landfills.**

**Figure 13.3**
In this example, acrylic is used with cotton to encourage deliberate areas of pilling only in the acrylic areas. A braid knitting pattern has been designed in the front using a cotton yarn. Since the cotton won't pill, wearing this garment over time will celebrate the pilling from the acrylic without making the garment look worn out.
*Amy Williams*

Chiyacat/Shutterstock

# 14

# Imitation Leather (Substitutes for 100 Percent Genuine Leather)

There are several types of alternatives to 100 percent genuine leather, including thermoplastic polyurethane (TPU), polyurethane laminate (PUL), and polyvinyl chloride (PVC). They are often used for jackets, handbags, shoes, and upholstery and were developed as inexpensive alternatives to leather. These materials are manufactured, synthetic products.[1] The following chapter will focus on TPU, PUL, and PVC and will also discuss "hybrid" forms of 100 percent genuine leather, including PU coated split leather and bonded leather. While these leather types are generated from pure leather, they are processed using synthetic materials that are bonded to the leather and cannot be later separated. This falls into the "imitation leather" category because it is a downgraded version of 100 percent genuine leather that imitates the real thing.

TPU, PUL, and PVC are derived from petroleum and are nonbiodegradable and not recyclable. This presents a problem because these materials tend to be used in inexpensive, fast-fashion garments and accessories that are designed to have a short life span. Efforts to look at alternative materials that could either be recyclable or biodegradable could influence the overall environmental impact that products using these materials have on the world around us.

**Table 14.1 Benefits and Potential Impacts**

| | Description | Benefit | Impacts |
|---|---|---|---|
| Thermoplastic polyurethane | • Heat bonding lamination process where no solvents are necessary.<br>• Two types of TPU are common: polyester based and polyether based. | • Can be waterproof and weigh less than genuine leather.<br>• Can be constantly reused, which is why it's often used for disposable diapers.<br>• Could be considered "animal friendly" since it is not derived from the hide of a cow.<br>• These products have the visual aesthetics of genuine leather, but at substantially less cost. | • Less durable than genuine leather.<br>• The base material used to form polyurethane compounds is a by-product of the oil refining process.<br>• Almost all commercial grade polyurethanes available are based on two isocyanates: TDI (toluene diisocyanate) and MDI (methylene bisdiphenyl diisocyanate).<br>• TDI is considered a volatile organic compound (VOC) and has acute and chronic effects on humans. |
| Polyurethane laminate | • A polyurethane coating is laminated onto fabrics such as polyester or cotton and uses solvents in a chemical bonding process. | • Durable, waterproof, flexible.<br>• Can be constantly reused.<br>• These products have the visual aesthetics of genuine leather, but at substantially less cost. | |
| Polyvinyl chloride | • Synthetic plastic polymer that can be made in rigid and flexible forms. | • Polyvinyl chloride is a versatile plastic that can take on a variety of characteristics—rigid, filmy, flexible, and leathery—with relatively limitless applications. | • Could be considered "animal friendly" since it is not derived from the hide of a cow.<br>• These products have the visual aesthetics of genuine leather, but at substantially less cost.<br>• Less durable than genuine leather.<br>• Dioxin (the most potent carcinogen known), ethylene dichloride, and vinyl chloride are emitted during the production of PVC and can cause acute and chronic health problems, including endocrine disruption reproductive and immune system damage,[2,3] and cancer.[4]<br>• Chemical stabilizers are necessary in the creation of PVC, including **lead**, **cadmium**, and **organotins**. Phthalates are used to soften PVC.[5] Certain phthalates have been banned from use in toys in the European Union (DINP, DIDP, and DNOP)[6] and the United States (the same phthalates and DEHP, BBP, and DBP)[7] and are known to cause acute and chronic health problems.[8]<br>• During use, dioxins and phthalates can leach, flake, or outgas from PVC over time, again emitting dioxin and heavy metals into the air and water.[9]<br>• Less breathable than polyurethane.<br>• Not biodegradable. |

**BOX 14.1**
**OTHER SUBSTITUTES FOR 100 PERCENT GENUINE LEATHER**

**PU Coated Split Leather**

**PU coated split leather**, also known as "**PU split**," comes from the same hide as 100 percent genuine leather. The hide is prepared and tanned. Since the hide is too thick to use on its own, it is split into layers. The top layer is of the highest quality and is considered pure leather. The lower layer, called "split," is also considered 100 percent genuine leather and looks like suede. For the processing of PU coated split leather, the tanner applies a thin layer of polyurethane (PU) with foil or extrusion that hardens on top. A hair cell pattern can be embossed on the PU layer so that it looks like genuine leather. PU coated split leather is not considered 100 percent leather.

**Bonded Leather**

Bonded leather also comes from the same hide as 100 percent genuine leather. The hide is prepared and tanned. Small pieces of leather that are cut away from the final usable piece are combined with composite materials and spread out in sheets. Foil is put on top to resemble the top layer of leather. Bonded leather is to 100 percent genuine leather as particle board is to wood. Bonded leather is not considered 100 percent leather.

- Visit **Chapter 26: Recycled/Circular Textiles Technologies** for more information about recyclability.
- Visit **Appendix: Biodegradability in the Fashion Fibers STUDIO** for information about designing for biodegradability.

# Optimize the Sustainability Benefits of Imitation Leather

- Promote suppliers who use alternatives to PVC.
- Promote suppliers who use water-based solvents for PUL.
- Investigate "vegan leathers" made out of polyester or polyamide microfiber, to allow them to be recyclable.

# Availability

- Vegetable, low- or no-chrome tanned leather is readily available.
- Water-based solvents for PUL are currently being researched.
- Microfiber made out of synthetics, such as polyester, are readily available.

# Fashion Applications

- Bonded leather is applicable for belts, shoe soles, and furniture upholstery; it could be creatively applied to bags.

**Figure 14.1**
Stella McCartney "Elyse" platform shoes made with imitation leather: 60 percent polyurethane, 40 percent polyester, and wood that comes from a certified sustainable source.
*Vanni Bassetti/Getty Images*

**BOX 14.2 SUSTAINABLE BENEFIT**

Stella McCartney is known for her ongoing commitment to animal-friendly fashion and uses only "vegan leathers" that have been created using highly skilled manufacturing techniques.

- Microfiber for leather substitutes made out of polyester are applicable for shoes, handbags, and upholstery.
- Microfibers for leather substitutes made out of 100 percent chemically recycled polyester are available through Toray in the United States.[10]

# Potential Marketing Opportunities

- **Post-consumer recycled leather:** If from used garments.
- **Vegan leather:** "Vegan leather" can be claimed if 100 percent of the materials are nonleather and animal free. *NOTE: Simply saying "vegan leather" is not enough to substantiate sustainability claims, since this material is generally derived from petroleum and can be toxic.*
- **100% recyclable:** If polyester or polyamide microfiber is used and infrastructure to collect products and garments is in place.

**Figure 14.2**
One hundred percent faux leather Stella McCartney handbag from spring/summer 2016 collection.
*Pascal Le Segretain/Getty Images*

**BOX 14.3 POTENTIAL MARKETING OPPORTUNITY**

"Vegan leather" can be claimed if 100 percent of the materials are nonleather and animal free.

# INNOVATION EXERCISES

## Design

1. Design a garment that emphasizes one or more of the sustainability benefits of one type of imitation leather.
2. Design a garment that circumvents one or more of the environmental impacts of one type of imitation leather.
3. Design a garment or accessory that considers what will happen at the end-of-use stage of the life cycle. The product should address longevity, recyclability, biodegradability, disassembly for reuse, etc.
4. Currently, imitation leather recycling is not commercially available. While this technology is being developed, think about how you could extend the use of products by redesigning them into new products. Design a new product or use imitation leather from collected garments and products.
5. Investigate innovative design opportunities that will allow the majority of imitation leather accessories to be recyclable, minimizing the load on landfills.
6. Research a supplier that produces a polyester alternative to imitation leather. Design a product that emphasizes the material characteristics and is recyclable.

## Innovation Exercise Example

**Investigate innovative design opportunities that will allow the majority of imitation leather accessories to be recyclable, minimizing the load on landfills.**

**Figure 14.3**
In this example, a 100 percent microfiber polyester cape is designed using laser cutting details and trims to allow for recyclability.
*Amy Williams*

## Merchandising

7. Create a fifty-word marketing communication message for consumers about the sustainability benefits of imitation leather.
8. Incorporating information from Chapters 8 and 14, develop a table that a retail buyer could use to quickly compare the advantages and disadvantages of leather and imitation/vegan leather.
9. Create a retail point-of-purchase sign or hangtag to promote fashion merchandise made from imitation leather.
10. Go online and find three fashion retailers who sell fashion merchandise made from imitation leather. What are the retailers and fashion merchandise offered? Why did they select this material for their fashion merchandise? How do they promote the advantages of this material to consumers? Do you believe their promotional strategies are accurate and effective? Why or why not?
11. Develop a marketing research protocol (e.g., focus group protocol, online survey, observational strategy) to determine consumers' perceptions of imitation leather for fashion merchandise.

TACrafts/iStock

# 15

# Polyethylene

Polyethylene (PE) is a synthetic, manufactured plastic that is the most commonly used polymer in the world. It constitutes about one-third of all plastics produced worldwide, and its applications for fashion and apparel are mainly shipping bags, packaging, and industrial uses.[1]

The main considerations with polyethylene are that it is used in abundance, is nonbiodegradable, and is generally disposed of. Many fashion companies have expressed a desire for a "better" solution for shipping apparel, but a better system has yet to be worked out. Efforts to decrease overall consumption and come up with a better solution for shipping could influence the environmental impact that polyethylene has on the world around us.

# Benefits

Polyethylene is tough, flexible, lightweight (some are featherweight), waterproof, and easy to process.

There are three main types of polyethylene: low-density polyethylene (LDPE), linear-low-density polyethylene (LLDPE), and high-density polyethylene (HDPE).

Products in these categories are used in diverse applications: bags for newspapers, dry-cleaning, and frozen foods; sandwich bags; shrink wrap; squeezable bottles; coatings on milk cartons and hot and cold beverage cups; lids; toys; flexible tubing; plastic grocery bags; retail shopping bags; milk jugs; juice, detergent, and household cleaner bottles; safety protective clothing; and apparel/product bags for shipping.

Polyethylene is inexpensive. It costs less than a penny to manufacture 140 grams.[2]

Polyethylene's sole monomer, ethylene, is relatively nontoxic to humans and the environment.

# Potential Impacts

### Processing

The manufacturing process for polyethylene requires nonrenewable resources and high water use. For the production of 1 kilogram of high-density polyethylene, 1.5 kilograms of fossil fuels are required and over 3 kilograms of water.[3]

**Figure 15.1**
Polyethylene's most substantial environmental impact is at its end-of-use stage. It is highly durable, but it is also used for "throwaway" products such as shipping bags for clothing.
*Nastco/iStock*

### End of Use

Polyethylene's most substantial environmental impact is at its end-of-use stage. Despite its durability (plastic bags can hold more than 100 times their weight!), polyethylene was not designed for longevity, but for immediate throwaway.

Carbon dioxide emissions are released when high-density polyethylene is incinerated. This could happen in countries that do not have access to more sophisticated disposal, recycling, and waste-to-energy methods.[4]

♦ Visit **Chapter 26: Recycled/Circular Textiles Technologies** for more information about recyclability.

♦ Visit **Appendix: Biodegradability in the Fashion Fibers STUDIO** for information about designing for biodegradability.

# Optimize Sustainability Benefits

- Encourage the use of bio-derived polyethylene. Bio-derived polyethylene is derived from renewable resources, such as sugarcane. Bio-plastics have a lower carbon footprint, and some are recyclable and **compostable**. There is no guarantee that they are manufactured with chemicals that are not toxic to the environment.
- Consider the use of compostable bio-plastic for shipping.
- Encourage the use of recycled polyethylene.

**Figure 15.2**
Compostable bio-plastic from Tipa Corp.
*Tipa Corp*

# Availability

- Recycled polyethylene is readily available globally.

# Applications

- Apparel shipping bags, shopping bags.

**Figure 15.3**
Currently, most apparel is shipped to retail stores in individually packaged PE bags that are generally not recycled. In this example, garments are shipped in small containers lined with compostable PLA material. Not only does this save resources, but it will save time for the receiving department to unpack the garments.
*Yelena Safranova*

# Potential Marketing Opportunities

- **X% recycled content:** Regulations require stating percentage recycled if not 100 percent recycled content.
- **X% plant-based material:** If verified and used.

# INNOVATION EXERCISES

### Design

1. Design a product that emphasizes one or more of the sustainability benefits of polyethylene.
2. Design a product that circumvents one or more of the environmental impacts of polyethylene.
3. Develop an innovative way to reuse polyethylene for garment shipping of apparel and products to stores.
4. Develop an idea for an alternative fiber or material to replace polyethylene bags for garment product shipping. What process or system would be needed to ensure that an alternative fiber works?
5. How would you reward customers for reusing bags?
6. Develop a 100 percent compostable shopping bag that biodegrades in less than two weeks.

### Merchandising

7. Create a fifty-word marketing communication message for fashion retail customers about the sustainability benefits of polyethylene.
8. Go online and find three companies that sell packaging used in the fashion industry made from polyethylene. What are the companies and what types of packaging do they sell? Why did they select this material for this type of packaging? How do they promote the advantages of this material to their customers? Do you believe their promotional strategies are accurate and effective? Why or why not?
9. Develop a take-back and recycling program for polyethylene packaging, including the objectives of the program and implementation strategies.

# REVIEW AND RECOMMENDATION

## Preferred

- Polyethylene is a preferred fiber for two main reasons: Polyethylene's sole monomer, ethylene, is relatively nontoxic to humans and the environment; it is also recyclable.

## Recommended Improvements

- Use recycled polyethylene.
- Develop end-of-use strategy to guarantee take-back and recyclability.

stocksnapper/iStock

16

# Polypropylene (PP)

Polypropylene (PP) is a long-chain synthetic polymer composed of at least 85 percent by weight of ethylene, propylene, or other olefin units. Polypropylene is a synthetic, manufactured polyolefin fiber that is used in a number of diverse applications ranging from carpet to technical and outdoor apparel to geotextiles and product packaging. Global polypropylene usage is at 2.6 million tons, and efforts to address sustainability innovations could make a significant impact on the industry and the planet.[1]

# Benefits

Polypropylene's characteristics have been perfected over the years since it was originally developed in the 1950s. It has excellent durability, strength, and resiliency while still being lightweight. Polypropylene has good resistance to ultraviolet degradation, stains, and spilling and excellent wicking action—which make this material great for carpets.[2] These features also eliminate the need for water- and stain-repellent finishes.

Polypropylene's natural buoyancy also makes it perfect for high-performance apparel such as wetsuits and swimsuits.

Polypropylene blends well with other fibers, and when used capitalizes on its excellent wicking properties.[3]

Dyeing is not possible with polypropylene fiber, and therefore it must be pigmented during manufacture. This process, known as "solution dyeing," can be done with all the synthetic fibers (nylon, PET, PE, and PP). It is the only option for polypropylene, however, as the fiber will not take up dyes. Since colors are incorporated during the fiber-forming stage, no dyeing is necessary—which means no pollution from dyeing.[4]

♦ Visit **Chapter 23: Dyeing and Printing** for more information on impacts associated with dyeing.

Polypropylene's low softening point encourages consumers to launder their products using low temperatures for washing and ironing, thereby minimizing water and energy use associated with consumer care and washing.[5]

Polypropylene is made up of a sole **monomer**, **propylene**, which is relatively nontoxic to humans and the environment.

# Potential Impacts

### Processing

Typical of synthetic fibers, production for polypropylene varies among manufacturers. Individual manufacturers have variations in their processes to achieve certain properties such as light stability and heat sensitivity.[7]

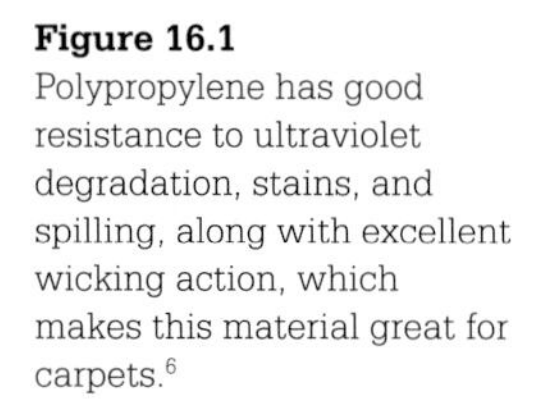

**Figure 16.1**
Polypropylene has good resistance to ultraviolet degradation, stains, and spilling, along with excellent wicking action, which makes this material great for carpets.[6]
*Jan Stromme/Getty Images*

The manufacturing process for polypropylene requires nonrenewable resources and high water and energy use.[8,9,10]

Fuel released by vehicles used to transport oil and waste causes pollution and $CO_2$ emissions.[11]

- Visit **Chapter 26: Recycled/Circular Textiles Technologies** for more information about recyclability.
- Visit **Appendix: Biodegradability in the Fashion Fibers STUDIO** for information about designing for biodegradability.

# Optimize Sustainability Benefits

- Promote the use of recycled polypropylene.
- Promote research on bio-derived polypropylene. Bio-derived polypropylene is derived from renewable resources, such as sugarcane. Bio-plastics have a lower carbon footprint, and some are recyclable and compostable. There is no guarantee that they are manufactured with chemicals that are not toxic to the environment. Also, plants used for bio-plastic feedstock can be grown with fertilizers and pesticides.[12,13]
- Promote OEKO-TEX® certified polypropylene.[14] OEKO-TEX® is an independent, third-party certifier that offers two certifications for textiles: OEKO-TEX® 100 (for products) and OEKO-TEX® 1000 (for production sites/factories). OEKO-TEX® 100 labels aim to ensure that products pose no risk to health. OEKO-TEX® certified products do not contain allergenic dyestuffs or dyestuffs that form carcinogenic aryl-amines. The certification process includes thorough testing for a long list of chemicals. Specifically banned are AZO dyes, carcinogenic and allergy-inducing dyes, pesticides, chlorinated phenols, extractable heavy metals, emissions of volatile components, and more.

**Figure 16.2**
Promote the use of recycled polypropylene.
*SSPL via Getty Images*

# Availability

- Recycled polypropylene is available from suppliers in Europe and China.
- Bio-derived polypropylene is currently an advancing technology and is not readily available.

# Fashion and Other Applications

- High-performance gear for backpacking/canoeing/mountain climbing, wetsuits, swimsuits, running/cycling clothing, inexpensive carpets, outdoor seating, upholstery, and industrial uses.

# Potential Marketing Opportunities

- **X% recycled content:** Regulations require stating percentage recycled if not 100 percent recycled content.
- **X% plant-based material:** If verified and used.

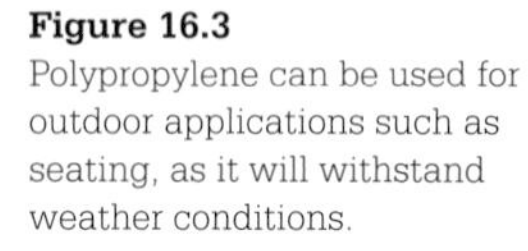

**Figure 16.3**
Polypropylene can be used for outdoor applications such as seating, as it will withstand weather conditions.
*Jamie Grill/Getty Images*

# INNOVATION OPPORTUNITIES

## Design

1. Design a garment that emphasizes one or more of the sustainability benefits of polypropylene.
2. Design a garment that circumvents one or more of the environmental impacts of polypropylene.
3. Consider what will happen to a polypropylene product at the end-of-use stage of the life cycle. Design a polypropylene product that addresses longevity, recyclability, biodegradability, disassembly for reuse, etc.
4. How would you create the infrastructure to ensure take-back of polypropylene products?
5. Design polypropylene products with reuse in mind in order to optimize resources embodied in the product.

## Merchandising

6. Create a fifty-word marketing communication message for consumers about the sustainability benefits of recycled polypropylene.
7. Incorporating information from chapters 6, 13, and 16, develop a table that a retail buyer could use to quickly compare the advantages and disadvantages of 100 percent wool, 100 percent acrylic, 100 percent polypropylene, and 100 percent recycled polypropylene.
8. Create a retail point-of-purchase sign or hangtag to promote fashion merchandise (including home fashions) made from recycled polypropylene.
9. Go online and find three fashion retailers that sell fashion merchandise (including home fashions) made from recycled polypropylene. What are the retailers and fashion merchandise offered? Why did they select this fiber for their fashion merchandise? How do they promote the advantages of this fiber to consumers? Do you believe their promotional strategies are accurate and effective? Why or why not?
10. Develop a marketing research protocol (e.g., focus group protocol, online survey, observational strategy) to determine consumers' perceptions of polypropylene and recycled polypropylene for fashion merchandise (including home fashions).
11. Develop a retail take-back and recycling program for polypropylene products—including the objectives of the program and implementation strategies.

# REVIEW AND RECOMMENDATION

## Preferred

Polypropylene is a preferred fiber for two main reasons: Polypropylene is made up of a sole monomer, propylene, which is relatively nontoxic to humans and the environment, and it is recyclable.

## Recommended Improvements

- Use recycled polypropylene.
- Incorporate environmental dyeing practices.
- Develop end-of-use strategy to guarantee take-back and recyclability.

*DNY59/iStock*

# 17

# Rayon/Viscose Made from Wood

Rayon is primarily made from wood and is categorized as a manufactured cellulosic fiber created from cellulose found in trees. It is typically derived from spruce, pine, or eucalyptus (although many plants or trees can be used to make rayon) and is then chemically processed and regenerated to form a new polymer using the rayon process. Bamboo can be processed in the same way to produce rayon/viscose and is detailed in **Chapter 18, Rayon/Viscose Made from Bamboo**. Rayon/viscose can also be made from cotton linters.

Although rayon is generally not considered a sustainable fiber due to its highly chemical process, new forms of rayon-type materials are emerging—such as lyocell—that have the same material characteristics but are produced by less-toxic processes and operate in a **closed-loop** system where outputs are recovered, filtered, and reused.

# Benefits

Rayon is the oldest manufactured fiber. The rayon process was developed in the late 1800s as an inexpensive alternative to silk. Rayon has a silklike aesthetic, drapes well, is easy to dye, and is highly absorbent. It is a good conductor of heat, so it is a cool, comfortable fiber good for use in warm weather.

Rayon is also relatively inexpensive compared to other fibers, and it blends well with many fibers—sometimes used to reduce cost, or contribute luster, softness, absorbency, or comfort.[1]

# Potential Impacts

## Cultivation

Wood feedstock may be sourced from ancient and endangered forests.[2] Approximately 100 million trees are logged annually for fabrics—approximately one-third of which are from ancient and endangered forests.[3] See Box 17.1 for information on the not-for-profit environmental organization Canopy, which is working with clothing brands to ensure rayon/viscose is not sourced from ancient and endangered forests by 2017.

**BOX 17.1**
**CANOPY WORKS WITH APPAREL COMPANIES TO ENSURE RESPONSIBLE FEEDSTOCK SOURCING**

Canopy (www.canopyplanet.org), a not-for-profit environmental organization, is working collaboratively with over sixty clothing brands and viscose producers as part of the CanopyStyle initiative to ensure that rayon/viscose is not sourced from ancient and endangered forests by 2017. Canopy is dedicated to protecting forests, species, and climate and has collaborated with more than 750 companies to develop innovative solutions, make their supply chains more sustainable, and help protect our world's remaining ancient and endangered forests. The organization's partners in the CanopyStyle campaign include H&M, Zara/Inditex, Stella McCartney, Levi Strauss & Co., EILEEN FISHER, and Topshop.

## Processing

To transform hardwood-derived materials into silky fabric, the cellulose must be separated from other compounds found in trees. Sodium hydroxide (caustic soda) and sodium sulfide are commonly used to remove the lignin that binds the wood fibers together, and in some cases bleach is required to whiten the pulp. In a complex process the pulp is steeped in caustic soda to produce alkali cellulose, which is then aged or oxidized before reacting with carbon disulfide to create sodium cellulose xanthate. This xanthate is dissolved in caustic soda to form a syrup-like spinning solution, or "rayon," which can then be extruded through a spinneret to form rayon fibers.[4]

The rayon manufacturing process is chemically intensive and requires copious amounts of water. Wastewater effluents from processing must be properly treated to avoid contamination of surrounding water bodies. Air emissions caused by the rayon process include sulfur, nitrous oxides, carbon disulfide, and **hydrogen disulfide**. Chronic exposure to carbon disulfide can cause damage to the nervous system in humans.[5]

## Consumer Care/Washing

Rayon is typically dry-clean only, due to the delicacy of the fabric when wet. Some types of rayon can be machine or hand washed.

Certain at-home detergents and commercial dry-cleaning chemicals have been reported to have detrimental effects on humans and the environment; they contribute to ozone depletion and can pollute wastewater.

- Visit **Appendix B: Consumer Care and Washing** for more information.
- Visit **Chapter 22: Bleaching** for information on the environmental impacts of bleaching.
- Visit **Chapter 23: Dyeing and Printing** for more information on impacts associated with dyeing.

- Visit **Chapter 26: Recycled/Circular Textiles Technologies** for more information about recyclability.
- Visit **Appendix: Biodegradability in the Fashion Fibers STUDIO** for information about designing for biodegradability.

# Optimize the Sustainability Benefits of Rayon

- Discourage suppliers from using trees from ancient and endangered forests, endangered species habitats, or areas logged in contravention of indigenous and community rights as feedstock for rayon fabrics, especially those harvested from the following endangered forest areas: Canadian Boreal Forest, including the Broadback Forest, Quebec; coastal temperate rainforests of British Columbia, Canada, and the Pacific Northwest United States; Russia's Taiga; Democratic Republic of Congo; and tropical forests of Indonesia and the Amazon.[6]
- Encourage suppliers to use raw materials sourced from responsibly managed forests outside of ancient and endangered forests certified to the **Forest Stewardship Council (FSC)** certification system.[7]

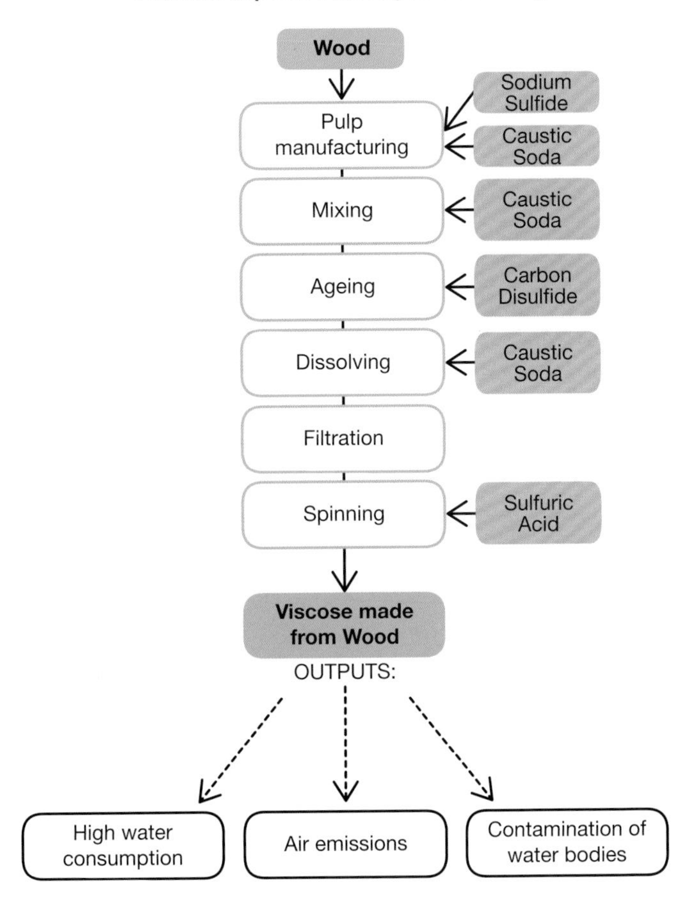


**Figure 17.1**
The traditional rayon/viscose manufacturing process is chemically intensive with high water consumption and emissions to air, and it has the potential to contaminate receiving water bodies if the discharge isn't treated.
*Fairchild Books*

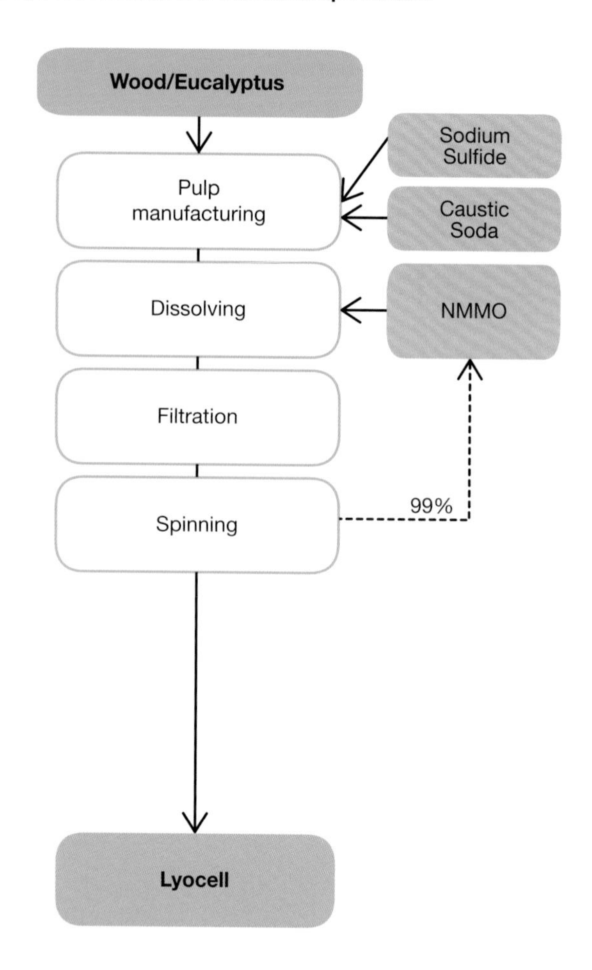


**Figure 17.2**
The lyocell process is a closed-loop process for producing rayon/viscose materials where most of the chemicals used are recovered, filtered, and reused.
*Fairchild Books*

- Use lyocell instead of traditional rayon/viscose to reduce the chemical impacts of the fabric. New forms of rayon-type materials are emerging and can be made through various types of processes, including lyocell. Lyocell material properties are similar to traditional rayon, and fiber production is also similar to that of generic rayon in that hardwood material is dissolved through an intensive chemical process into a pulp, which is then extruded to form fibers. However, for lyocell, the solvent used to transform the pulp into fiber is amine oxide (NMMO=N-Methyl-Morpholine-N-Oxide), which is considered nontoxic. The lyocell fiber manufacturing process also operates as a closed loop system, in which 99 percent of the solvent is recovered, filtered, and reused. Any remaining emissions are broken down harmlessly in biological wastewater treatment plants.[8]

# Availability

A few suppliers in China are currently offering wood pulp for rayon from FSC-certified forests. Expressing interest in a supplier's fiber sourcing, and requesting that the supplier does not source from ancient and endangered forests and instead sources from FSC-certified sources or other alternative fibers (such as bamboo from degraded lands or recycled fabrics) can influence the supplier's raw material sourcing strategy and lead to greater availability of responsibly sourced feedstock for rayon fabric.

OEKO-TEX® Standard 100 certified rayon is available. Manufacturers can be found at https://www.oeko-tex.com.

# Fashion Applications

Rayon and lyocell can be used in a variety of textile woven and knitted applications. Depending on the weight and construction of the cloth, these fabrics may be suitable for shirts, skirts, dresses, evening gowns, home furnishings, and bedding.

**Figure 17.3**
Use lyocell instead of traditional rayon/viscose to reduce the chemical impacts of the fabric. Lyocell can be used in a variety of textile applications and weights, including shirts, skirts, and dresses.
*Pascal Le Segretain/Getty Images*

# Potential Marketing Opportunities

- **Lyocell process:** If used and you have verified through your fabric supplier. Communicate the sustainable benefits to your customers through point-of-sale (POS) marketing, hangtags, and website.
- **FSC certified:** Must be verified as FSC certified and can be claimed on POS items at retail.
- **OEKO-TEX® Standard 100 certified:** If verified and used.

# INNOVATION EXERCISES

### Design

1. Design a garment that emphasizes one or more of the sustainability benefits of rayon/viscose made from wood.
2. Design a garment to circumvent one or more of the environmental impacts of rayon/viscose made from wood.
3. Design a collection of garments that are 100 percent biodegradable. Consider trims and dyes.
4. Create a garment where the design influences the customer to hand wash the garment.
5. Design a garment using a fabric that is an appropriate alternative to rayon/viscose. Explain why your fabric choice is more desirable from an ecological perspective.
6. Contact not-for-profit Canopy (www.canopyplanet.org) to learn more about the impacts of rayon/viscose sourcing on ancient and endangered forests, and what you and/or your brand can do to ensure that you are not sourcing from ancient and endangered forests or other controversial sources.

### Merchandising

7. Create a fifty-word marketing communication message for consumers about the sustainability benefits of rayon made from wood using the lyocell process.
8. Incorporating information from chapters 1, 3, 7, 10, and 17, develop a table that a retail buyer could use to quickly compare the advantages and disadvantages of 100 percent conventional cotton, 100 percent organic cotton, 100 percent bamboo linen, 100 percent silk, 100 percent recycled polyester, and 100 percent lyocell rayon made from wood.
9. Create a retail point-of-purchase sign or hangtag to promote fashion merchandise made from 100 percent lyocell rayon made from wood.
10. Go online and find three fashion retailers who sell fashion merchandise made from 100 percent lyocell rayon. What are the retailers and fashion merchandise offered? Why did they select this fiber for their fashion merchandise? How do they promote the advantages of this fiber to consumers? Do you believe their promotional strategies are accurate and effective? Why or why not?
11. Develop a marketing research protocol (e.g., focus group protocol, online survey, observational strategy) to determine consumers' perceptions of lyocell rayon for fashion merchandise.

# REVIEW AND RECOMMENDATION

### Preferred

Rayon/viscose made from wood is preferred only if:

- Bamboo from FSC-certified plantations has been used for the pulp, and preferably bamboo on degraded lands.
- It has been processed with the lyocell process.

### Recommended Improvements

- Encourage use of rayon/viscose from recycled fiber when available or from straw and other nonwood fibers.
- Encourage development of recycling processes for rayon/viscose.
- Incorporate environmental dyeing practices.
- Develop end-of-use strategy to guarantee take-back and recyclability.

# OTHER SOURCES

1. Frühwald 1995

richwai777/iStock

18

# Rayon/Viscose Made from Bamboo

Rayon/viscose made from bamboo is categorized as a manufactured cellulosic fiber created from cellulose found in the bamboo plant. It is derived from bamboo, which is then chemically processed and regenerated to form a new polymer using the viscose process.

# Benefits

Bamboo is a "rapidly renewable" resource, meaning that it grows quickly and can be harvested at least once a year.[1]

Bamboo is a biologically efficient, low-maintenance crop that requires few chemical inputs during the growing season. It is mainly rain fed and can grow in diverse climates.

Due to its speedy growth and little input needed for growing, some say that using bamboo as an alternative to slower-growing wood trees could help slow deforestation.[2]

Viscose from bamboo drapes well, is easy to dye, and is highly absorbent. It is a good conductor of heat, so it is a cool, comfortable fiber good for use in warm weather. Viscose made from bamboo is prized for its softness and comfort.

**Table 18.1 Fast-Growing Renewable Fibers**

| Fiber | Timing | Length |
|---|---|---|
| Bamboo (for linen or viscose) | 40 days[3] | 24 meters |
| Hemp | 3 months[4] | 4 meters |
| Jute | 3–4 months[5] | 1–4 meters |
| Flax | 3–4 months[6] | 1 meter[7] |

# Potential Impacts

### Processing

Some species of bamboo are highly invasive, meaning they take over natural vegetation.

To transform plant-derived materials into silky fabric, the cellulose must be separated from other compounds found in bamboo. Sodium hydroxide (caustic soda) and sodium sulfide are commonly used to remove the lignin that binds the plant fibers together and in some cases bleach is required to whiten the pulp. In a complex process, the pulp is steeped in caustic soda to produce alkali cellulose, which is then aged or oxidized before reacting with carbon disulfide to create sodium cellulose xanthate. This xanthate is dissolved in caustic soda to form a syrup-like spinning solution or "viscose," which can then be extruded through a spinneret to form viscose fibers.[8]

The viscose manufacturing process is chemically intensive and requires copious amounts of water. Wastewater effluents from processing must be properly treated to avoid contamination of surrounding water bodies. Air emissions caused by the viscose process include sulfur, nitrous oxides, carbon disulfide, and hydrogen disulfide. Chronic exposure to carbon disulfide can cause damage to the nervous system in humans.[9]

The rayon (made from bamboo) manufacturing process is chemically intensive and requires copious amounts of water. Wastewater effluents from

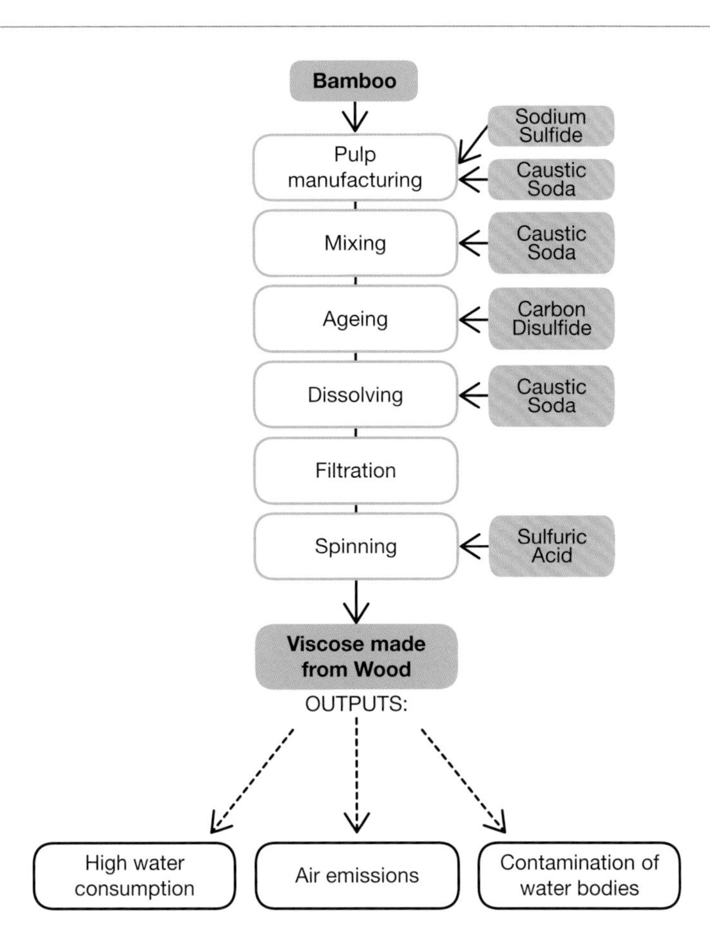


**Figure 18.1**
The traditional rayon/viscose (made from bamboo) manufacturing process is chemically intensive with high water consumption and emissions to air, and it has the potential to contaminate receiving water bodies if the discharge isn't treated.
*Fairchild Books*

processing must be properly treated to avoid contamination of surrounding water bodies.

## Consumer Care/Washing

Rayon/viscose made from bamboo is typically dry-clean only, due to the delicacy of the fabric when wet. Some types of viscose can be machine or hand washed. Certain at-home detergents and commercial dry-cleaning chemicals have been reported to have detrimental effects on humans and the environment; they contribute to ozone depletion and can pollute wastewater.

- Visit **Appendix B: Consumer Care and Washing** for more information.
- Visit **Chapter 22: Bleaching** for information on the environmental impacts of bleaching.
- Visit **Chapter 23: Dyeing and Printing** for more information on impacts associated with dyeing.
- Visit **Chapter 26: Recycled/Circular Textiles Technologies** for more information about recyclability.
- Visit **Appendix: Biodegradability in the Fashion Fibers STUDIO** for information about designing for biodegradability.

# Optimize Sustainability Benefits

- Know the difference between linen made from bamboo and viscose made from bamboo. Viscose made from bamboo employs a chemically intensive process and has high environmental and social impacts due to emissions to air and water during processing.
    - Visit **Chapter 3: Bamboo Linen** for more information about linen made from bamboo.
- Encourage suppliers to use raw materials sourced from **Programme for the Endorsement of Forest Certification (PEFC)** and **Forest Stewardship Council (FSC)** certified plantations.
- Investigate viscose processing methods that use enzymes instead of chemicals.
- New forms of viscose-type materials are emerging and can be made through various types of processes including lyocell. Lyocell fiber production is similar to that of generic viscose in that bamboo material is dissolved through an intensive chemical process into a pulp, which is then extruded to form fibers. However, for lyocell, the solvent used to transform the pulp into fiber is amine oxide (NMMO=N-Methyl-Morpholine-N-Oxide), which is considered nontoxic. The lyocell fiber manufacturing process also operates as a closed-loop system, in which 99 percent of the solvent is recovered,

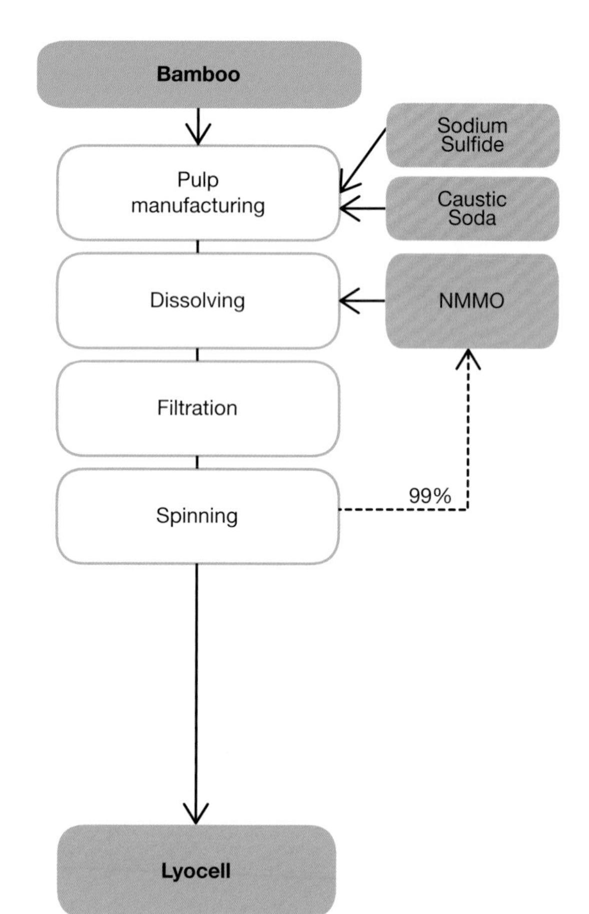


**Figure 18.2**
The lyocell process is a closed-loop process for producing rayon/viscose materials where most of the chemicals used are recovered, filtered, and reused.
*Fairchild Books*

**BOX 18.1**
**SUSTAINABLE BENEFIT TO LYOCELL MANUFACTURING PROCESS**

The manufacturing process for lyocell is a closed-loop process, in which most of the chemicals used for processing into fiber are recovered, filtered, and reused.

filtered, and reused. Any remaining emissions are broken down harmlessly in biological wastewater treatment plants.[10]

- EcoPlanet Bamboo is industrializing bamboo as a sustainable alternative fiber for the textile industry. The company is working to combine a certified source of fiber with innovative closed-loop pulping technology to transform the supply chain of major clothing brands globally. This model begins with the restoration of degraded land into high-yielding bamboo farms that also provide significant biodiversity and social impact.

# Availability

Few suppliers are currently offering viscose made from bamboo from PEFC- and FSC-certified plantations. Expressing interest in PEFC and FSC certification can influence the supplier's raw material sourcing strategy and lead to greater availability of responsibly sourced feedstock for viscose made from bamboo.

Texplan, a Spanish company, is working with suppliers in China to produce lyocell fiber made from bamboo that is also FSC certified.[11]

Litrax, a Swiss company, has developed a process that uses enzymes instead of chemicals for processing bamboo into viscose.[12]

# Fashion Applications

Rayon/viscose made from bamboo and lyocell made from bamboo can be used in a variety of textile woven and knitted applications. Depending on the weight and construction of the cloth, these fabrics may be suitable for shirts, skirts, dresses, evening gowns, home furnishings, and bedding.

**Figure 18.3**
Expressing interest in PEFC and FSC certification can influence the supplier's raw material sourcing strategy and lead to greater availability of responsibly sourced feedstock for viscose made from bamboo.
*Camille Rebelo/EcoPlanet Bamboo*

# Potential Marketing Opportunities

- **Viscose:** Bamboo is being marketed strongly as an eco-friendly fiber. While the raw material is a rapidly renewable natural resource, viscose made from bamboo fabric employs a highly pollutive process in its manufacture. The European Commission has issued a directive on textile names, 2008/121/EC. This directive states how textile products should be marketed and sold in the EU. The name "bamboo" does not appear in this directive; therefore, it cannot be used for the purposes of compulsory description of fiber composition. The name "viscose" is included in this directive and should be used to describe the fibers corresponding to the definition "regenerated cellulose fiber obtained by the viscose process for filament and discontinuous fiber." This includes viscose made from bamboo fibers. This should be done consistently on labeling, hangtags, and point-of-sale (POS) marketing.[13]
- **Bamboo lyocell:** If bamboo is processed with a lyocell process.
- **Bamboo from FSC-certified plantations:** Using bamboo for viscose from FSC-certified plantations is best, and if verified, it can be claimed on POS items at retail.
- **Fast-growing natural resource:** Bamboo is considered a fast-growing resource, as seen in Table 18.1, and can be marketed as such.
- **Low water footprint in cultivation:** Bamboo is biologically efficient, is mainly rain fed, and can grow in diverse climates.

Viscose made from bamboo is the most mismarketed fiber on the market, when it comes to sustainability claims. As a result, there have been several regulations created to clarify these misleading claims:

- The European Commission has issued a directive on textile names, 2008/121/EC, and the name "bamboo" does not appear in this directive. The name "viscose" is included in this directive and should be used to describe the fibers corresponding to viscose made from bamboo fibers.[14]
- The **Federal Trade Commission (FTC)** in the United States also requires that fabric should not be referred to as "bamboo." In the United States, the bamboo textile should be referred to more specifically as "viscose made from bamboo."[15]

# INNOVATION OPPORTUNITIES

## Design

1. Design a garment that emphasizes one or more of the sustainability benefits of rayon/viscose made from bamboo.
2. Design a garment that circumvents one or more of the environmental impacts of rayon/viscose made from bamboo.
3. Design a collection of garments that are 100 percent biodegradable. Consider trims and dyes.
4. Create a garment where the design influences the customer to hand wash the garment.
5. Design a garment using a fabric that is an appropriate alternative to rayon/viscose. Explain why your fabric choice is more desirable from an ecological perspective.

**BOX 18.2 BEWARE OF MISLEADING CLAIMS ABOUT BAMBOO RAYON**

Rayon/viscose from bamboo is often mismarketed and claimed as "eco-friendly."

### Merchandising

6. Create a fifty-word marketing communication message for consumers about the sustainability benefits of rayon made from bamboo using the lyocell process.
7. Incorporating information from chapters 3, 7, 10, 17, and 18, develop a table that a retail buyer could use to quickly compare the advantages and disadvantages of 100 percent bamboo linen, 100 percent silk, 100 percent recycled polyester, 100 percent rayon made from wood, 100 percent rayon made from bamboo, and 100 percent rayon made from bamboo using the lyocell process.
8. Create a retail point-of-purchase sign or hangtag to promote fashion merchandise made from 100 percent lyocell rayon made from bamboo.
9. Go online and find three fashion retailers who sell fashion merchandise made from 100 percent lyocell rayon from bamboo. What are the retailers and fashion merchandise offered? Why did they select this fiber for their fashion merchandise? How do they promote the advantages of this fiber to consumers? Do you believe their promotional strategies are accurate and effective? Why or why not?
10. Develop a marketing research protocol (e.g., focus group protocol, online survey, observational strategy) to determine consumers' perceptions of lyocell rayon from bamboo for fashion merchandise.

## REVIEW AND RECOMMENDATION

### Preferred

Rayon/viscose made from bamboo is preferred only if:

- Bamboo from FSC-certified plantations has been used for the pulp, and preferably bamboo on degraded lands.
- It has been processed with the lyocell process.

### Recommended Improvements

- Encourage use of rayon/viscose recycled fiber when available.
- Encourage development of recycling processes for rayon/viscose.
- Incorporate environmental dyeing practices.
- Develop end-of-use strategy to guarantee take-back and recyclability.

AlexTois/iStock

# 19

# Lyocell

Lyocell fiber is made from cellulose originating primarily from eucalyptus wood, and it has unique material properties that can be suitable for a variety of applications. Like rayon, lyocell begins with a wood pulp, but the manufacturing process for lyocell is vastly different: it is a closed-loop process, in which 99.8 percent of the chemicals used for processing into fiber are recovered, filtered, and reused. This feature makes it a viable, sustainable alternative to cotton, viscose, and possibly other synthetics.

A majority of the world's lyocell comes from Lenzing. TENCEL® lyocell is the registered brand name for lyocell fibers manufactured by Lenzing in Austria.

# Benefits

Eucalyptus trees, from which lyocell is derived, grow rapidly on marginal lands without artificial irrigation, gene manipulation, or synthetic pesticides. Lenzing claims that the trees used as feedstock for TENCEL® lyocell are harvested from sustainably managed farms certified by the Forest Stewardship Council (FSC) or Programme for the Endorsement of Forest Certification (PEFC).[1] Lenzing has also confirmed that pulp used for the manufacture of TENCEL® lyocell fiber is supplied from production locations that comply with the EU Timber Regulation.[2,3]

To transform hard wood into lyocell fabric, the cellulose must be separated from other compounds found in the trees. The wood material is dissolved through an intensive chemical process into a pulp, which is then extruded to form fibers. The solvent used to transform the pulp into fiber is amine oxide (NMMO=N-Methyl-Morpholine-N-Oxide), which is considered nontoxic.[4] The TENCEL® lyocell fiber manufacturing process also operates as a closed-loop system, in which 99.8 percent of the solvent is recovered, filtered, and reused, and any remaining emissions are broken down harmlessly in biological wastewater treatment plants.[5,6]

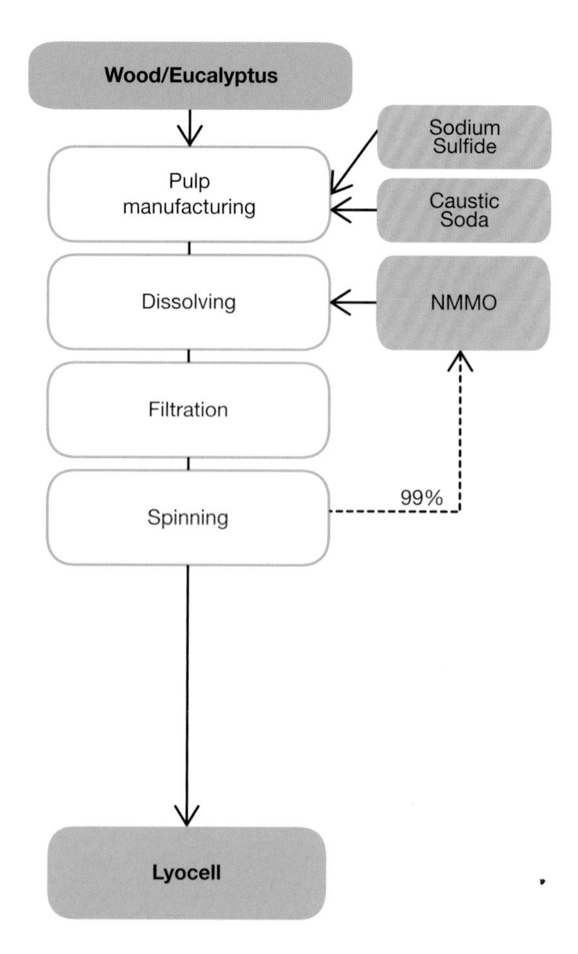


**Figure 19.1**
The lyocell process is a closed-loop process for producing man-made cellulosic materials, where most of the chemicals used are recovered, filtered, and reused.
*Fairchild Books*

Lyocell has a smooth fiber surface and round cross-section. This fibrillar structure enables improved dye pickup and achieves vibrant colors and a slight sheen on the surface of the fabric while using less dyestuff and less water throughout the dyeing process.

In all applications, TENCEL® lyocell has moisture wicking properties and good absorbency, is wrinkle resistant, and has good drapability. High wet and dry strength make it an excellent fiber for denim applications.

The dyeing process for TENCEL® lyocell can significantly reduce water consumption and dye and chemical use due to its good color absorption.

Since TENCEL® lyocell fiber absorbs and redirects moisture (i.e., sweat) fewer washings may be needed, resulting in water and energy savings as well as reduced wear and tear that occurs with repeated laundering. This should be communicated to the customer.

# Potential Impacts

Currently, the solvents used for TENCEL® lyocell are derived from petrochemicals. However, the solvents are being recovered and reused.

- Visit **Chapter: 23 Dyeing and Printing** for more information on impacts associated with dyeing.
- Visit **Chapter 26: Recycled/Circular Textiles Technologies** for more information about recyclability.
- Visit **Appendix B: Consumer Care and Washing** for more information.
- Visit **Appendix: Biodegradability in the Fashion Fibers STUDIO** for information about designing for biodegradability.

**Figure 19.2**
TENCEL® lyocell fabric.
*SSPL/Getty Images*

# Optimize Sustainability Benefits

- Ensure that lyocell fabric is TENCEL® lyocell from Lenzing because this company discloses its raw material source, which is verified as coming from certified sustainably harvested forests.
- Cross-check with Lenzing that the fabric supplier is one of its supply chain partners: branding.lenzing.com.
- Ask for Lenzing "certification" on the fabric—this is readily available if TENCEL® lyocell content is 30 percent or above. Exceptions are made for certain products like denim with cotton warps.

# Availability

TENCEL® lyocell is readily available, and the fiber is produced in four production sites: Mobile, Alabama, United States; Grimsby, United Kingdom; Heiligenkreuz, Burgenland, Austria; and Lenzing, Austria.

# Fashion Applications

Suitable product applications for lyocell and TENCEL® lyocell depend on the fabric weight and construction, and include denim, dress pants, jackets, dress shirts, blouses, active wear, sleepwear, work wear, and home textiles in woven and knit applications.

# Potential Marketing Opportunities

- **TENCEL® lyocell:** Fabric should be referred to as TENCEL® lyocell.
- **FSC- or PEFC-certified:** If TENCEL® lyocell from FSC- or PEFC-certified plantations is used, then this can be claimed.
- **Biodegradable:** All fibers, yarns, trims, and dyes used to manufacture the product or garment must also be biodegradable or disassembled before disposal. This should be substantiated with documentation that the product can completely break down into nontoxic material by being processed in a facility where compost is accepted. A secondary label or marketing material should be provided to instruct customer.

**Figure 19.3**
Dress made from TENCEL® lyocell from Los Angeles–based apparel company The Reformation.

*The Reformation*

**Figure 19.4**
Dress made from TENCEL®
Lyocell from Patagonia.
*Courtesy of Patagonia*

# INNOVATION OPPORTUNITIES

## Design

1. Design a garment that emphasizes one or more of the sustainability benefits of TENCEL® lyocell.
2. Design a garment that circumvents one or more of the environmental impacts of TENCEL® lyocell.
3. Design a collection of garments that are 100 percent biodegradable. Consider trims and dyes.
4. Create a garment where the design influences the customer to hand wash the garment.
5. Design a garment with lyocell used in high-perspiration areas of the garment to take advantage of its moisture absorbing and wicking properties.
6. Explore innovative fabrications that use blends with lyocell (organic cotton/recycled cotton, for example). Design a garment using one of these blends.

7. Design garments and products with reusable (synthetic) trims and a biodegradable body. Design the product so that nonbiodegradable trims, tags, buttons, etc., can be easily separated from the main body of the product at the end of use.

### Merchandising

8. Create a fifty-word marketing communication message for consumers about the sustainability benefits of TENCEL® Lyocell.
9. Incorporating information from chapters 7, 10, 18, and 19, develop a table that a retail buyer could use to quickly compare the advantages and disadvantages of 100 percent silk, 100 percent recycled polyester, 100 percent Lyocell rayon made from bamboo, and 100 percent TENCEL® Lyocell.
10. Create a retail point-of-purchase sign or hangtag to promote fashion merchandise made from 100 percent Lyocell.
11. Go online and find three fashion retailers who sell fashion merchandise made from 100 percent TENCEL® Lyocell. What are the retailers and fashion merchandise offered? Why did they select this fiber for their fashion merchandise? How do they promote the advantages of this fiber to consumers? Do you believe their promotional strategies are accurate and effective? Why or why not?
12. Develop a marketing research protocol (e.g., focus group protocol, online survey, observational strategy) to determine consumers' perceptions of TENCEL® Lyocell for fashion merchandise.

## REVIEW AND RECOMMENDATION

### Preferred

TENCEL® lyocell is preferred fiber because:

- The manufacturing process for TENCEL® lyocell is a closed-loop process, in which most of the chemicals used for processing into fiber are recovered, filtered, and reused.
- Lenzing also claims that the trees used as feedstock for TENCEL® lyocell are harvested from sustainably managed farms certified by the Forest Stewardship Council or Programme for the Endorsement of Forest Certification.[7]
- TENCEL® lyocell is biodegradable and potentially recyclable.

### Recommended Improvements

- Ensure feedstock comes from FSC- or PEFC-certified wood.
- Encourage use of lyocell and TENCEL® lyocell recycled fiber when available.
- Encourage development of recycling processes for lyocell and TENCEL® lyocell.
- Incorporate environmental dyeing practices.
- Develop end-of-use strategy to guarantee take-back and recyclability.

*malerapaso/iStock*

# 20

# Modal

Modal's material properties are similar to traditional viscose; however it has three major differentiating characteristics: 1) the manufacturing process for modal is less toxic than traditional rayon/viscose processing, 2) the beech trees aren't harvested from endangered forests, and 3) it has greater wet strength, which means that modal is machine washable.

A majority of the world's modal comes from Lenzing. Lenzing Modal® is the registered brand name for the second generation of viscose fibers manufactured by Lenzing in Austria at the only fully integrated pulp and fiber production site in the world. It is categorized as a manufactured fiber created from cellulose originating primarily from beech trees. It is then chemically processed and regenerated to form a new polymer using a process similar to viscose.

# Benefits

Beech trees are soil enhancers, breed naturally, do not need artificial irrigation, and are indigenous to the region around Austria. Though not entirely immune, the beech tree is naturally resistant to pests and disease.[1] Lenzing also states that the yield per acre of Lenzing Modal® is six times higher than cotton yield, and its cultivation requires considerably less water.[1]

The modal manufacturing process is nontoxic and operates in a system where 95 percent of outputs are recovered, filtered, and reused.[2]

Modal's distinguishing fiber characteristics are its extra softness and high wet strength. Modal is machine washable and can be washed in cold water and line dried, thereby minimizing water and energy use associated with consumer care and washing. Modal fibers do not shrink or get pulled out of shape when wet, like viscose.[3]

Modal drapes well, is easy to dye, and is highly absorbent.

Lenzing Modal® fiber is harvested from beech trees in the PEFC- (Programme for the Endorsement of Forest Certification) certified European forests.

Lenzing Modal® COLOR is a unique process that addresses environmental impacts from dying. Lenzing Modal® COLOR adds the color pigment directly to the fiber before the fiber is extruded, so dyeing is no longer necessary. Tests show when processing Lenzing Modal® COLOR, up to 80 percent of energy and up to 76 percent of water can be saved in comparison to the standard fiber (in jet dyeing).[4] Lenzing Modal® COLOR can achieve a range of eight colors with more in development.

Lenzing Modal® and Lenzing Viscose® Austria are the only man-made fibers that are **carbon neutral**.[5]

# Potential Impacts

### Dyeing and Finishing

Not all colors can be Lenzing Modal® COLOR, so traditional dyeing methods might have to be used.

- Visit **Chapter 22: Bleaching** for information on the environmental impacts of bleaching.
- Visit **Chapter 23: Dyeing and Printing** for more information on impacts associated with dyeing.

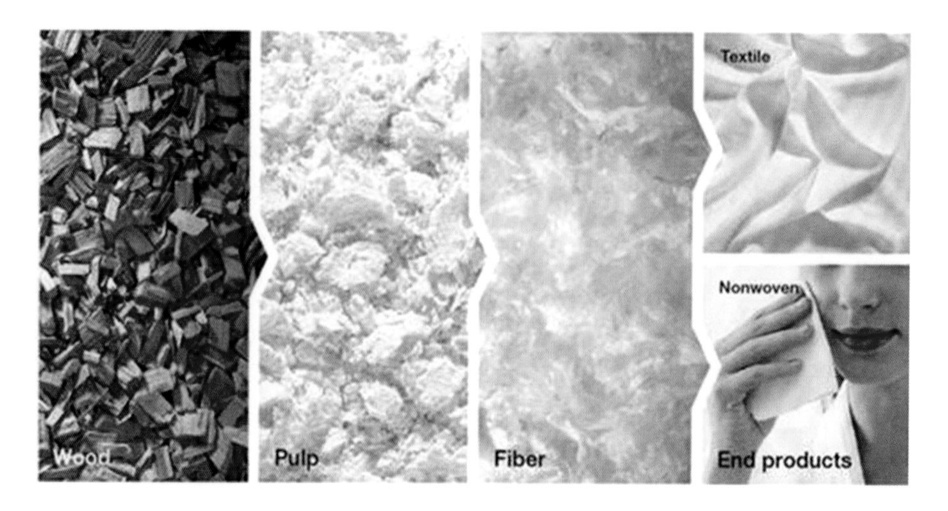


**Figure 20.1** Process for making Lenzing Modal® fiber.
*Lenzing Group*

- ♦ Visit **Chapter 26: Recycled/Circular Textiles Technologies** for more information about recyclability.
- ♦ Visit **Appendix B: Consumer Care and Washing** for more information.
- ♦ Visit **Appendix: Biodegradability in the Fashion Fibers STUDIO** for information about designing for biodegradability.

# Optimize Sustainability Benefits

- Ensure that modal fabric is from Lenzing, since this company discloses its raw material source, which is verified as coming from certified sustainably harvested forests.
- Cross-check with Lenzing that the fabric supplier is one of its supply chain partners.
- Ask for Lenzing "certification" on the fabric.
- Use Lenzing Modal® COLOR.
- Look for OEKO-TEX® certification:[6] OEKO-TEX® is an independent, third-party certifier that offers two certifications for textiles: OEKO-TEX® 100 (for products) and OEKO-TEX® 1000 (for production sites/factories). OEKO-TEX® 100 labels aim to ensure that products pose no risk to health. OEKO-TEX® certified products do not contain allergenic dyestuffs and dyestuffs that form carcinogenic aryl-amines. The certification process includes thorough testing for a long list of chemicals. Specifically banned are AZO dyes, carcinogenic and allergy-inducing dyes, pesticides, chlorinated phenols, extractable heavy metals, emissions of volatile components, and more.

  ♦ Visit **Part 3: Processing** of this text for more information about dyeing.

**BOX 20.1 FASHION APPLICATIONS FOR MODAL**

Modal fibers are especially favored for loungewear and sleepwear due to their softness and ability to be machine washed.

# Availability

- Lenzing Modal® is readily available, and the fiber is produced in Heiligenkreuz, Burgenland, Austria.
- Lenzing Modal® is often blended with cotton, wool, or synthetic fibers and allows easy tone-in-tone dyeing.

# Fashion Applications

Modal fibers are especially favored for loungewear, sleepwear, and undergarments. They can also be found in 100 percent and blends with cotton, wool, silk, and synthetic in apparel, outerwear, and household furnishings.[7]

**Figure 20.2**
Sleepwear made from modal jersey by Snoa Sleepwear.
*Dropship News*

# Potential Marketing Opportunities

- **FSC certified and PEFC certified:** Must be verified, and can be claimed on point-of-sale (POS) items at retail.
- **OEKO-TEX® Standard 100 certified:** If verified and used.

# INNOVATION EXERCISES

## Design

1. Design a garment that emphasizes one or more of the sustainability benefits of modal.
2. Design a garment that circumvents one or more of the environmental impacts of modal.
3. Design a garment that encourages hand washing in cold water. How would you communicate this to the customer?
4. Design a modal garment (or sleepwear collection for extra credit!) that emphasizes its extra softness and the fiber's ability to drape well on the body.
5. Develop an innovative way to explain to your customers the difference between the viscose process and the production process used for modal.
6. Create a collection of garments using modal and the color palette available through Lenzing Modal® COLOR.
7. Design a garment from modal that is 100 percent biodegradable.

## Merchandising

8. Create a fifty-word marketing communication message for consumers about the sustainability benefits of modal.
9. Incorporating information from chapters 7, 10, 18, 19, and 20, develop a table that a retail buyer could use to quickly compare the advantages and disadvantages of 100 percent silk, 100 percent recycled polyester, 100 percent lyocell rayon made from bamboo, 100 percent TENCEL® lyocell, and 100 percent modal.
10. Create a retail point-of-purchase sign or hangtag to promote fashion merchandise made from 100 percent modal.
11. Go online and find three fashion retailers who sell fashion merchandise made from 100 percent modal. What are the retailers and fashion merchandise offered? Why did they select this fiber for their fashion merchandise? How do they promote the advantages of this fiber to consumers? Do you believe their promotional strategies are accurate and effective? Why or why not?
12. Develop a marketing research protocol (e.g., focus group protocol, online survey, observational strategy) to determine consumers' perceptions of modal for fashion merchandise.

# REVIEW AND RECOMMENDATION

## Preferred

Modal is definitely a preferred fiber! Lenzing Modal® has the benefit of being created by a company with the intention of carrying sustainability throughout its entire process. Not only does modal have exceptional fiber characteristics, but it also does not use trees from ancient or endangered forests and it has been developed with environmentally friendly production processes in place. Modal's only limitation is that it is best for a limited number of fashion applications.

## Innovation Exercise Example

**Design a modal garment that emphasizes its extra softness and the fiber's ability to drape well on the body.**

**Figure 20.3**
In this design example, modal is used for a robe that emphasizes its ability to drape well when wrapped around the body.
*Amy Williams*

*kyoshino/iStock*

# 21

# Azlon (from Soy)

Azlon is a fiber that is composed of regenerated, naturally occurring protein.[1] Azlon (from soy) is derived from soybean, a renewable raw material resource. Azlon (from soy) fabric has many desirable features that may offer an alternative to the conventional fiber industries: cashmere, cotton, polyester, nylon, rayon, linen, and wool.

# Sustainable Benefits

Commercial production of azlon (from soy) fiber involves bioengineering techniques. The soybean protein's structure is modified through the use of enzymes and by incorporating polyvinyl alcohol (PVA) either into the spinning solution or as a central core surrounded by an outer coating of soy protein. This process adds strength and wearability characteristics to the fiber. The agents used in the processing of soy fiber are said to be nontoxic and it's claimed that waste can be used as animal feed once the protein has been extracted.

Azlon (from soy) fiber in its pure form is biodegradable; however, the use of finishes and PVA likely reduces this potential.

**Figure 21.1**
The design in this example shows a Baby Kimono Body suit from Babysoy, 50 percent organic cotton and 50 percent azlon (from soy).
*Babysoy*

**Figure 21.2**
The agents used in the processing of soy fiber are nontoxic and safe to wear against children's skin.
*Johner Images/Getty Images*

**Table 21.1 Azlon (from Soy) Fiber Offers Features That Are Comparable, and in Some Cases Superior to, Other Fibers[2]**

| Feature | Qualities | Compares to | Superior to |
|---|---|---|---|
| Luster | Silky luster | Silk | n/a |
| Drape | Excellent drape | Silk | n/a |
| Comfort/softness | Light and soft, airy | Cotton, silk, cashmere | n/a |
| Absorbency | Good moisture absorption | Cotton | n/a |
| Strength | High breaking strength | Wool, cotton, silk | Wool, cotton, silk |
| Good colorfastness | Stable in bright light and perspiration conditions | Silk | Silk |
| Easy care | Anticrease, easy wash, fast drying | Cotton | Silk |

# Potential Impacts

## Cultivation

Large-scale industrial azlon (from soy) farming is fertilizer and pesticide intense. Many of these chemicals are used by farmers in developing countries where education, access to information, and understanding of the dangers posed by hazardous chemicals are lacking. The use of toxic chemicals can also diminish soil fertility over time, and irrigation runoff can pollute surrounding water bodies if not treated.

**Figure 21.3**
Large-scale industrial soy farming is fertilizer and pesticide intense. Many of these chemicals are used by farmers in developing countries where education, access to information, and understanding of the dangers posed by hazardous chemicals are lacking.
*Johannes Kroemer/Getty Images*

In addition, soy is one of the top three largest genetically modified crops and is linked to a number of impacts including increased herbicide use (see genetic modification [GM] in the cotton section). In 2014, 94 percent of all soy in the United States was genetically modified (compared to 96 percent of cotton crops).[3] GM is a relatively new technology, and its long-term effects are not yet fully understood.

### Bleaching and Dyeing

The natural color of azlon (from soy) fiber is light yellow and is difficult to remove. The use of bleach is therefore recommended, and when very light colors are desired, whitening agents are also suggested. Chlorine bleach can form halogenated organic compounds in the wastewater. These compounds bioaccumulate in the food chain, are known teratogens and mutagens, are suspected human carcinogens, and can cause reproductive harm.

- Visit **Chapter 22: Bleaching** for information on the environmental impacts of bleaching.
- Visit **Chapter 23: Dyeing and Printing** for more information on impacts associated with dyeing.
- Visit **Chapter 26: Recycled/Circular Textiles Technologies** for more information about recyclability.
- Visit **Appendix B: Consumer Care and Washing** for more information.
- Visit **Appendix: Biodegradability in the Fashion Fibers STUDIO** for information about designing for biodegradability.

## Optimize Sustainability Benefits

- Promote the use of azlon from organic soybeans.
- When blended with natural fibers, azlon (from soy) fiber can enhance positive aspects and reduce negative impacts of other fibers.

**Table 21.2 Blend Azlon (from Soy) with Other Natural Fibers to Enhance Positive Aspects of Each Fiber**

| Blend | Percentage | Features |
|---|---|---|
| Azlon (from soy)/ cashmere | (80/20) | Enhances the hand and lowers the manufacturing cost of cashmere products. Could potentially lower the destructive impact of overgrazing by cashmere goats. Gives superior luster and comfort, as well as antipilling and drape properties. |
| Azlon (from soy)/wool | (50/50) | Optimizes the luster, soft hand, and strength of azlon (from soy) fiber, and the elasticity and heat-retaining properties of wool. Enables spinning of high-count yarn. |
| Azlon (from soy)/silk | (50/50) | Improves resistance to staining from perspiration and water; improves lightfastness and moisture permeability. Enables higher quality yarns and fabrics at lower cost. |
| Azlon (from soy)/ cotton | (50/50) | Enhances the comfort, luster, moisture-permeability, quick-dry, and drape properties of cotton. Upgrades cotton products.[4] |

# Availability

Soy protein fiber, yarn, and fabric are available from Harvest SPF Textile Co. Ltd. in China.

# Fashion Applications

Azalon (from soy) is ideal for products that are worn close to the skin, such as intimates, sleepwear, activewear, and infants'/children's clothing. It can also be used in fine gauge knitwear.

# Potential Marketing Opportunities

- **Azlon (from soy):** Fabric should not be referred to simply as "soy" since soy is not considered a fiber. The fiber must be referred to as "azlon (from soy)."
- **Renewable natural resource:** FTC requires substantiation for any general environmental benefit claim.
- **Antimicrobial:** Though soy fibers are generally promoted as having natural antibacterial properties, antimicrobial agents must be registered with the Environmental Protection Agency (EPA) for the specific purpose for which they are used, and specific language must be used on labeling.
- **Organic:** Some soy fabrics are marketed with organic certification. Organic certification requires that the soy crop be grown without the use of disallowed synthetic chemicals and uses nongenetically modified seed. Certified organic farming of soy must be monitored and certified by an independent third party. Ask for proof of the organic certification.
- **Biodegradable (depending on dyes and trims used):** Documentation is required to substantiate that the product can completely break down into nontoxic material through "normal routes of disposal" within one year. "Normal routes of disposal" cannot be landfill but must be processed in a facility where compost is accepted. A secondary label or marketing material has to be provided to instruct customer.
- **Non-chlorine bleached:** If alternative bleach is used.
- **Non–genetically modified:** If grown organically, or seed purchase receipts are provided.

# INNOVATION EXERCISES

### Design

1. Design a garment that emphasizes one or more of the sustainability benefits of azlon (from soy).
2. Design a garment that circumvents one or more of the environmental impacts of azlon (from soy).
3. Design a garment using the natural golden color of azlon (from soy) fiber to avoid dyeing and associated pollution impacts.

4. Create a garment using a blend of azlon (from soy) with another natural fiber in order to enhance positive aspects and reduce negative impacts. Explain why you chose the blend.
5. Consider blending azlon (from soy) with silk, wool, or cotton to maximize positive aspects and reduce negative impacts of these fibers.
6. Design a garment where azlon (from soy) is strategically placed on the garment where perspiration is most likely to build up, to take advantage of the fiber's absorbency characteristics.
7. Design a garment where azlon (from soy) is strategically placed on the garment in areas that are most likely to take more wear and tear, to take advantage of the fiber's physical durability.
8. Design a baby's or children's garment where azlon (from soy) is used.

### Merchandising

9. Create a fifty-word marketing communication message for consumers about the sustainability benefits of azlon from soy.
10. Incorporating information from chapters 7, 10, 18, 19, 20, and 21, develop a table that a retail buyer could use to quickly compare the advantages and disadvantages of 100 percent silk, 100 percent recycled polyester, 100 percent lyocell rayon made from bamboo, 100 percent TENCEL® lyocell, 100 percent modal, and 100 percent azlon (from soy).
11. Create a retail point-of-purchase sign or hangtag to promote fashion merchandise made from 100 percent azlon (from soy).
12. Go online and find three fashion retailers who sell fashion merchandise made from 100 percent azlon (from soy). What are the retailers and fashion merchandise offered? Why did they select this fiber for their fashion merchandise? How do they promote the advantages of this fiber to consumers? Do you believe their promotional strategies are accurate and effective? Why or why not?
13. Develop a marketing research protocol (e.g., focus group protocol, online survey, observational strategy) to determine consumers' perceptions of azlon (from soy) for fashion merchandise.

## REVIEW AND RECOMMENDATION

### Preferred

Azlon (from soy) fiber is preferred for use as a sustainable fiber because:

- Azlon (from soy) fiber is derived from a natural renewable source.
- Azlon (from soy) fiber is biodegradable.
- Azlon (from soy) fiber has performance qualities equal, and sometimes superior, to other natural fibers. This fiber can be used as an alternative to fibers with greater environmental impacts.

### Recommended Improvements

- Use azlon (from soy) fabric that is derived from organically grown soybeans.
- Incorporate environmental dyeing practices.
- Develop end-of-use strategy (for compostability or recyclability).

# OTHER SOURCES

1. www.newfibers.com.tw/pub/LIT_8.asp?ctyp=LITERATURE&pcatid=0&catid=2794
2. Natural Fibres, Soy Protein Fibre n.d.
3. Swicofil, Soybean protein fiber properties 2013

Jevtic/iStock

# FUTURE FIBERS: MANUFACTURED FIBERS

This Future Fibers part closer presents innovative alternatives to traditional manufactured fibers. The companies profiled here have all presented unique ways of circumventing challenges and impacts associated with processing and cultivating virgin manufactured fibers.

## Polylactide

**Polylactide (PLA)** is mainly made from sugars derived from corn, though any abundantly available sugar, such as wheat, sugar beets, or sugarcane could also be used. PLA is a new class of polymer that is biodegradable under optimum conditions. Ingeo from NatureWorks LLC is a readily available brand name of PLA.[1]

**Figure P2.1**
Polylactide is a bio-based fiber that is derived from corn.
*Reuben Schulz/iStock*

### Sustainable Benefits

Polylactide has excellent resiliency, outstanding crimp retention, and good wicking ability. It has good thermal insulation and breathability, high UV protection, and excellent hand and drape.[2] PLA has natural resistance to staining and low odor retention, and it can be machine washed and dried, with no need to iron.

### Biodegradability/Recyclability

Polylactide is fully biodegradable as long as component parts of the garment are also made from PLA and optimum composting conditions are present. When used in combination with nonrenewables, PLA cannot be claimed as biodegradable. PLA will not biodegrade in landfills. It requires a balance of oxygen, moisture, aeration, and steady temperatures of 49–60°C—a balance that is typically found in industrial composting facilities. Home compost heaps do not provide the required combination of temperature and humidity to trigger decomposition.

Furthermore, PLA *cannot* go into the regular recycling bin and can contaminate a batch of polyethylene terephthalate (PET). There is currently no standard system for differentiating PLA plastics.

### Availability

PLA is commercially available as yarn and fiber from Ingeo from NatureWorks LLC and can be produced at large quantities.

### Fashion Applications

Pillows, comforters, mattress pads, performance activewear, fashion apparel, outdoor furniture, and nonwovens, such as diapers.

**Figure P2.2**
DuPont has commercialized polytrimethylene terephtalate (PTT) into a fiber called Sorona®, which can be used as an alternative to nylon 6.
*Sorona® is a registered trademark of E.I. du Pont de Nemours and Company. Image courtesy of the DuPont Company.*

## Polytrimethylene Terephtalate (Sorona® from DuPont[3])

Polytrimethylene terephtalate (PTT) is a bio-based fiber that can be used as an alternative to nylon 6. DuPont has commercialized the fiber into a product called Sorona®.

### Sustainable Benefits

PTT is a resilient fiber, is dyeable, and has outstanding elastic recovery.[4]

PTT can be easily blended with other fibers, including cotton, linen, wool, nylon, and polyester.

Sorona® uses up to 37 percent by weight of dextrose (sugar) from corn combined through a proprietary process with traditional petroleum-based feedstock and terephthalic acid. DuPont claims that producing Sorona® uses 30 percent less energy and releases 63 percent less greenhouse gas emissions compared to the production of nylon 6.

## Potential Impacts

Sorona® is **bio-based** and not biodegradable, which means that 37 percent of the fiber is derived from agriculture, rather than oil. When selecting fiber, it is always important to consider a fiber's biodegradability or recyclability, so it will not end up in a landfill after use.

While Sorona® can blend well with other natural and manufactured fibers, it is also important to note that when natural fibers are blended with Sorona®, they are no longer biodegradable, and when manufactured fibers are blended with Sorona® they are no longer recyclable. This could potentially lead to these garments again ending up in a landfill.

## Biodegradability/Recyclability

Due to the nonrenewable source of terephthalic acid, Sonora® is not considered a fully renewable fiber.[5] It is not biodegradable.[6] Recyclability is unknown.

- ♦ Visit **Appendix: Biodegradability in the Fashion Fibers STUDIO** for information about designing for biodegradability.

## Availability

Sorona®, a brand name for polytrimethylene terephtalate, is readily available from DuPont.

## Fashion Applications

PTT can be used for swimwear, activewear, and outerwear.

## Optimize Sustainability Benefits

- Ensure that these fibers you use are also addressing other potential impacts through the cultivation and processing of the feedstock and resulting fiber.
- If you are using a fiber that is under 100 percent bio-based, ensure there is potential for recyclability, since the fiber is likely not biodegradable.
- Using a biodegradable fiber is one thing, but ensuring it gets biodegraded is another. Since PLA is biodegradable only under optimum conditions that are typically available in an industrial composting facility, develop instructions for your customer on your website, point-of sale (POS) materi-als, or hangtags letting them know how to dispose of their garment.

## Potential Marketing Opportunities

- **Fast-growing natural resource:** If you have used a fast-growing biologi-cal resource as the feedstock.
- **X% bio-based:** Make it very clear what percentage of the fiber is biological.
- **X% recycled content:** If recycled content source has been verified.
- **Biodegradable:** All fibers, yarns, trims, and dyes used to manufacture the product or garment must also be biodegradable or disassembled before disposal. This should be substantiated with documentation that the prod-uct can completely break down into nontoxic material by being processed in a facility where compost is accepted. A secondary label or marketing material should be provided to instruct the customer.
- **Recyclable:** If there is a strategy in place to collect discarded textiles and clothing and recycle them into virgin materials.

### Consider Using These Future Fibers

| Fiber | Overview | Considerations | Availability | Applications |
|---|---|---|---|---|
| Bionic DPX (bionicyarn.com)<br>*Bionic* | Bionic DPX® is offered in two composition options: a dual stable fiber construction that blends recycled PET fibers with natural fibers to form a yarn with a soft texture, and a recycled PET yarn (100% rPET). To enhance functionality, DPX® can be spun around a high-tenacity filament or stretch filament core. Using this fiber avoids the need to use virgin resources. | DPX® blended textile PET/natural fiber is currently not recyclable; however the company is working on developing technology to recycle discarded Bionic DPX® textile into a next-generation fiber. See **Part 4: Circular Textiles** for information on potential recyclability and **Appendix: Biodegradability in the Fashion Fibers STUDIO** for information about designing for biodegradability. | Bionic works directly with potential partners to determine the feasibility of a partnership. Contact Bionic directly with interest: contact@bionicyarn.com. | DPX® is suitable for both knit and woven fabric applications: accessories, apparel, children's and baby's wear, denim, footwear, luxury/high-end, outdoor/athletic, outerwear, sleepwear and intimates, swimwear, and workwear. |
| ECONYL (econyl.com)<br>*Jason Ross Savage* | Econyl uses a system it calls the "ECONYL® Regeneration System" to turn recycled carpets, clothing, and fishing nets into virgin-quality raw material for nylon. Using this fiber avoids the need to use virgin resources. | Sustainable finishing treatments applied to the fabric for water repellency should also be considered. | Econyl works directly with potential partners to determine the feasibility of a partnership. | Applications include swimwear, accessories, hosiery, lingerie, sportswear, and outdoor apparel. |
| Apinat Bioplastics (apinatbio.com)<br>*Reuben Schulz/iStock* | APINAT is a biodegradable plastic that can be used for shoe soles. PUMA has used APINAT for the sole of its InCycle sneaker. | Having a biodegradable sole for a shoe is a good start. To ensure biodegradability, a take-back system and a strategy for successful composting must be in place, in addition to an innovative design to disassemble the shoe from the sole. | Commercially available. | Applicable for shoe soles. |

# INNOVATION EXERCISES

### Design

1. Design a garment that emphasizes one or more of the sustainability benefits of a manufactured future fiber.
2. Identify a potential environmental impact of one of these fibers. How would an intelligent design circumvent this impact?
3. Develop a collection that is entirely designed to be easily deconstructed to enable a take-back and recycling program. Experiment with seaming and a variety of disassembly mechanisms in different fabrics.

### Merchandising

4. Create a fifty-word marketing communication message for consumers about the sustainability benefits of polylactide or polytrimethylene terephtalate.
5. Incorporating information from chapters 10, 11, and 22, develop a table that a retail buyer could use to quickly compare the advantages and disadvantages of 100 percent polyester, 100 percent nylon, and 100 percent bio-based fiber (polylactide or polytrimethylene terephtalate).
6. Create a retail point-of-purchase sign or hangtag to promote fashion merchandise made from polylactide or polytrimethylene terephtalate.
7. Go online and find three fashion retailers who sell fashion merchandise made from polylactide or polytrimethylene terephtalate. What are the retailers and fashion merchandise offered? Why did they select this fiber for their fashion merchandise? How do they promote the advantages of this fiber to consumers? Do you believe their promotional strategies are accurate and effective? Why or why not?
8. Develop a marketing research protocol (e.g., focus group protocol, online survey, observational strategy) to determine consumers' perceptions of a bio-based fiber (polylactide or polytrimethylene terephtalate) for fashion merchandise.

## Review and Recommendation

### Recommended Improvements

- Incorporate environmental dyeing practices.
- Develop end-of-use strategy (whether compostability or recyclability).

*Mark Von Holden/Getty Images*

**Part 3**

# PROCESSING
## Overview

These days, garments are almost always dyed or printed, bleached, or washed in order to give them a more desirable, aged, or unique look. Processing is also the most often ignored when it comes to considering the environmental impacts of fashion design . . . which is understandable, given that sometimes these processes and finishes account for an overall low percentage by weight of the final garment. What might be surprising is that processing methods can produce some pretty significant negative ecological impacts to the planet and the communities surrounding the factories that process our clothes. These impacts can include negatively impacting receiving water bodies, harming both the people who rely on this water for drinking water and the aquatic plants and animals that live in these now-hostile environments. Some of the processes detailed in this section can have persistent bioaccumulative effects on the environment. Processing should not be ignored when it comes to considering the overall influence a garment will have on the environment.

The following chapters detail the impacts of bleaching, dyeing and printing, finishing, and garment washing and the recommended techniques and alternatives to mitigate or circumvent these potential impacts.

At the start of the design process, consider how these garment processes could impact the environment. Use this book to help you make better decisions and to develop sustainability innovations for garment processing that not only are unique but also improve the quality of the world around us.

Antonio de Moraes Barros Filho/WireImage/Getty Images

# 22

# Bleaching

Bleaching is the process of removing color. It is typically used to whiten cloth made from natural fibers where naturally occurring pigments are present. Bleaching is often used in conjunction with other cleaning processes such as desizing or scouring as preparatory steps for subsequent dyeing, printing, and/or finishing processes, since most of these processes work best on clean, white fabric. Fabric or garments can also be bleached, finished, and sold white.

Bleach is also effective at removing color from previously dyed/printed products and is commonly applied in industrial laundries to achieve "vintage" or "faded" looks (e.g., bleaching of indigo-dyed denim garments).

# Types of Bleaching Processes

## Hydrogen Peroxide

About 90 percent of fabric bleaching carried out prior to dyeing/printing is performed using hydrogen peroxide, a relatively safe chemical that decomposes into water and oxygen. Hydrogen peroxide is economical and readily available.

### Potential Impacts

Hydrogen peroxide bleaching requires high temperatures (above 60°C), which makes it an energy-intensive process. In addition, hydrogen peroxide is relatively unstable and requires chemical stabilizers and additional additives to optimize the bleaching process. These can be pollutive if left untreated and discharged into wastewater.[1]

## Chlorine Derivatives

About 90 percent of garment bleaching, used mainly for fading pre-dyed and preprinted items, is performed using chlorine derivatives (usually sodium or calcium hypochlorite) or potassium permanganate.

Hypochlorite is reactive enough to work well at cooler temperatures and breaks down into table salt, oxygen, and water. Like hypochlorite, potassium permanganate (PP) is a strong oxidizing agent. PP is particularly suitable for selective, or precision, applications (e.g., spraying or sponging) to achieve patterned effects, since its deep purple color makes it easy for the operator to see in real time and to visually track its application to the garment.

### Potential Impacts

If hypochlorite is exposed to organic material before it breaks down, it can react with that material to form halogenated organic compounds (organochlorines). Halogenated organic compounds are persistent, toxic compounds; may bioaccumulate in the food chain; are known teratogens/mutagens and suspected human carcinogens; and may cause reproductive harm. Hypochlorite can also react with acids, ammonia, and even dirt particles to form toxic gases.

Due to its highly toxic nature, personal protective equipment is required for workers handling PP.

Besides stripping color from the garments to which they are applied, chlorine derivatives and other oxidizing agents also attack and weaken the fiber and are particularly destructive to elastic fiber (spandex) in stretch fabrics, rendering the finished garment more fragile and susceptible to being discarded by the consumer more quickly.

# Alternative Technologies

## Ozone

Ozone gas is composed of three oxygen atoms ($O_3$) and is a much more powerful oxidizing agent than chlorine. It is very effective at fading pre-dyed/preprinted fabrics and garments, can be used at lower temperatures than hydrogen peroxide, and uses no water at all. In addition, ozone is completely biodegradable, as it reverts rapidly back to oxygen ($O_2$), leaving no chemical residue. However, the ozone bleaching process requires an alkali addition to swell the fiber and remove **motes** (small, immature seeds that may remain attached to the cotton fibers) from cotton, and scouring and sequestering agents still need to be used.[2] Additionally, ozone is highly unstable and cannot be stored. It must therefore be generated on-site as needed, meaning that any manufacturing facility intending to use ozone as a bleaching agent must invest in ozone-generating equipment.

## Laccase

Laccase is an enzyme that has proven effective at decolorizing or fading pre-dyed/preprinted apparel products. Generally speaking, enzymes are considered low impact because they are biodegradable (no waste products) and typically work well at low temperatures, thereby minimizing energy consumption. A small amount of enzyme often saves significant amounts of water, energy, and chemicals used in bleaching.[3] For example, Gentle Power Bleach™, a bleaching technology developed by Huntsman Textile Effects and Dupont™, is an enzyme-based peroxide process that enables pH neutral bleaching of textiles at 65°C. The company claims that using Gentle Power Bleach™ lowers treatment and rinsing temperatures and requires fewer rinse baths, resulting in significant water and energy consumption savings and eliminating the need for caustic soda altogether.[4]

Laccase's effectiveness varies depending on the specific dyestuff(s) used to initially dye/print the fabric or garment, but it works quite well on indigo-dyed denim products. Moreover, bleaching with laccase affects only the dyestuff and will not weaken the fiber or fabric as chlorine bleach does. However, enzymes are often more expensive than chemical options for bleaching, and some of the more popular denim shades may be difficult to achieve through enzyme technologies. For example, heavily faded denim effects cannot be replicated using laccase, and laccase often imparts a gray cast to indigo-dyed fabrics.

**Figure 22.1**
Range of colors that can be achieved by ozone treatment. Pictured here is denim treated with ozone from Saitex in Vietnam.
*Saitex International*

**Table 22.1 Optimize the Sustainability Benefits of Bleaching**

| Opportunity | Considerations |
|---|---|
| • Look for design opportunities to avoid bleaching. Darker, duller shades may not require bleaching prior to dyeing/finishing. | • Lighter, brighter shades do need to be chlorine bleached before dyeing.<br>• Grayed pastels may not need chlorine bleaching. |
| • Promote suppliers who use ozone bleaching processes or encourage ozone bleaching with existing suppliers.<br>• Promote the aesthetic of ozone bleach effects. | • Ozone has limited availability and is relatively expensive since it requires investment in an ozone generator.<br>• Ozone produces a different aesthetic than chlorine derivative or PP bleaching. |
| • Promote suppliers who use enzyme process bleaches.<br>• Promote the aesthetic of enzyme bleach effects. | • Enzyme produces a different aesthetic than chlorine derivative or PP bleaching. |
| • Promote suppliers who use low-temperature peroxide bleach processes, such as pad-batch systems, or encourage these processes with existing suppliers. | • Pad-batch bleaching is much slower than high-temperature peroxide bleaching and requires significant floor space to store batches for long periods of time. |
| • Avoid chlorine bleaching as much as possible. | • Peroxide might be slightly more expensive than chlorine-derivative bleaches, but it is less toxic. |
| • Promote proper wastewater treatment with your suppliers. | • Develop standards for laundries and dyehouses. |

**Figure 22.2**
Look for design opportunities to avoid bleaching. Darker, duller shades may not require bleaching prior to dyeing/finishing.
*Mark Von Holden/Getty Images*

# Availability

- Hydrogen peroxide is widely available globally.
- Chlorine derivatives and PP are used by the majority of garment laundries to fade pre-dyed/preprinted garments.
- Ozone is relatively expensive, and the equipment is not yet widely available.[5]
- Enzyme bleaching technologies are readily available globally.

# Fashion Applications

Not all fabrics require bleaching before they are dyed and finished. As a rule, only fabrics that contain naturally occurring pigments (such as hemp and linen) require bleaching with chlorine. Cotton fiber is naturally cream colored, and hydrogen peroxide bleaches will suffice in preparing cotton fabrics for dye. Wool and silk fibers yellow with the use of chlorine bleaches. Intended shade is another factor that determines the necessity of bleaching. Many

**Figure 22.3**
California Cloth Foundry uses a collaborative-development approach, working directly with farms, ranches, and the entire textiles/apparel supply chain to provide custom 100 percent farm-to-fashion nontoxic fabrications and fashion. This example of its natural-colored blends contains only all-natural finishes and uses no dyes, bleaches, or toxins.
*California Cloth Foundry*

darker, duller shades (dark brown, navy, black, etc.) are relatively easy to achieve on unbleached substrates, whereas lighter and brighter shades require a whiter base.

Hydrogen peroxide, ozone, and laccase bleaches can be effective substitutes for chlorine derivates and PP depending on the desired degree of fading, the desired shade/cast, the specific colorants involved, the equipment available, and the application technique(s) employed.

# Potential Marketing Opportunities

- **Water conservation (gallons of water/jean saved):** When ozone bleaching is used.
- **Nonchlorine bleached:** If alternative bleach is used.
- **Alternatives for bleaching:** Educational information on the corporate/brand website (perhaps in the same place as the corporate social responsibility [CSR] reports) could detail alternatives used for garment bleaching.

# INNOVATION EXERCISES

## Design

1. Design a garment that emphasizes one or more of the sustainability benefits of alternatives to chlorine bleaching.
2. Design a garment that circumvents one or more of the environmental impacts of bleaching.
3. Showcase an alternative bleaching method on a plain T-shirt.
4. Develop graphic symbols for hydrogen peroxide, ozone, laccase/enzyme, etc., to communicate with the customer.

## Merchandising

5. Create a fifty-word marketing communication message for consumers about the sustainability benefits of bleaching and its alternatives.
6. Create a retail point-of-purchase sign or hangtag to promote fashion merchandise made from an alternative to bleaching.
7. Go online and find three fashion retailers who sell fashion merchandise made with a bleaching and/or bleaching alternative. What are the retailers and fashion merchandise offered? Why did they select this finish for their fashion merchandise? How do they promote the advantages of this finish to consumers? Do you believe their promotional strategies are accurate and effective? Why or why not?
8. Develop a marketing research protocol (e.g., focus group protocol, online survey, observational strategy) to determine consumers' perceptions of bleaching alternatives for fashion merchandise.

# REVIEW AND RECOMMENDATION

## Not Preferred

Chlorine derivatives such as hypochlorite are not preferred. If hypochlorite is exposed to organic material before it breaks down, it can react with that material to form halogenated organic compounds (organochlorines). Halogenated organic compounds are persistent, toxic compounds; may bioaccumulate in the food chain; are known teratogens/mutagens and suspected human carcinogens; and may cause reproductive harm. Hypochlorite can also react with acids, ammonia, and even dirt particles to form toxic gases.

### Preferred

Hydrogen peroxide is preferred as long as the chemical stabilizers are safe and water discharge has been treated.

### Recommended Improvements

- Promote proper wastewater treatment with your suppliers.
- Encourage laccase or ozone treatments or other alternatives to traditional bleaching with chlorine.

wsfurlan/iStock

# 23

# Dyeing and Printing

The goal of adding color to a textile substrate is to produce an appealing, level, fast color on a product at a reasonable cost while providing choices to the customer.

Color is a critically important part of a fabric or garment and one of the single most important factors in the appeal and marketability of an apparel product, particularly during the "interest" phase of the consumer purchasing decision. An inappropriate or unattractive color may make a garment unmarketable no matter what the quality of the fiber, the yarn, the weave or knit, or the finish. Conversely, a poor-quality fabric may achieve big seller status purely because of its color. Aside from its attractiveness, color permanence (i.e., fastness) is important; most problems that consumers have with textile and apparel products are associated with color fading, bleeding, or staining (crocking).

At one time, colorants were limited to natural dyes and pigments obtained from plants, insects, and minerals. The first synthetic dye was developed in 1856, and by the early twentieth century, a wide range of synthetic dyes were readily available. Today, the textile and apparel industry uses over 700,000 tons of colorant annually, almost all of which is manufactured synthetically.[1]

Adding color to a textile substrate is a complex process. Slight differences in fabric caused by minor irregularities in fiber, yarn, fabric, or finishing can result in obvious color variations in finished products.[2] Fiber chemistry also plays an important role. A match between the chemistry of the colorant and that of the fiber is necessary in order for the color to be permanent. In addition, during use, any textile may be exposed to a wide variety of potential color degradants such as detergent, perspiration, dry-cleaning solvents, sunlight, or makeup. To achieve a durable color, the colorant must be attached to or trapped within the fiber by using a combination of heat, pressure, and chemicals. Fiber crystallinity, chemical finishes, and the fabric and yarn structures are all factors that influence the success of dyeing and printing.

# Overview of Environmental Concerns in Dyeing and Printing

- Formaldehyde—common ingredient of dispersing agents (for vat, sulfur, and disperse dyes), printing pastes (ingredient of resins added to promote cross-linking between binders and fibers), and colorant fixatives.
- Heavy metals—found in dyestuffs, dyeing auxiliary chemicals, and print pastes (as PVC stabilizers); often an unintended contaminant found in numerous chemicals.
- PVC and phthalates—used in plastisol printing pastes.
- Residual color in wastewater—due to poor exhaustion and/or fixation of colorants.
- Salt—used to promote exhaustion of reactive and direct dyes onto cotton substrates.
- Volatile organic compounds (VOCs)—in print pastes (particularly solvent based).
- High biological oxygen demand (BOD) or chemical oxygen demand (COD)—caused by substances in the wastewater after dyeing and finishing. BOD and COD create environments that are hostile to aquatic plants and animals.

**BOX 23.1**
**POTENTIAL IMPACT—UNTREATED WATER DISCHARGE FROM THE DYE PROCESS**

In developing countries such as India, Bangladesh, and China, dye water from textile mills often gets released to receiving water bodies without proper treatment. Generally, all inputs into dye processes end up as outputs in the water discharge. This affects local drinking water and the health of the surrounding communities.

**Figure 23.1**
Pollution from local textile dyeing mills in the Turag River in Bangladesh.
*Probal Rashid/LightRocket via Getty Images*

# Types of Dyes

- **Vat dye:** A dye that is insoluble in water and is reduced with a basic solution in order to activate the dye. Vat dyes are primarily used on cellulosic fibers due to the extremely high pH, but not exclusively.
- **Reactive dye:** A dye that attaches to the fiber chemically. Reactive dyes are primarily used on natural fibers (cotton, silk, wool, etc.) and rayon.
- **Disperse dye:** A dye that consists of microscopic non-water-soluble dye molecules dispersed in water that is then heated to 130°C in order to penetrate polyester fiber.
- **Acid dye:** A dye that bonds to a fiber via an ionic bond. Acid dyes are used on protein fibers such as silk and wool and may also be used on nylon.
- **Basic dye:** A cationic stain that will react with negatively charged material. They are not water soluble unless the base is converted to a salt. Basic dyes are strongly colored and bright but are not lightfast.
- **Sulfur dye:** A water-insoluble dye used for cotton with high toxicity. They are available in brown, black, and dark blue.
- **Naphthol dye:** A true cold water dye, naphthol dyes may be used in ice water. They are used to dye cellulosic fibers and silk. They are very popular for multicolored fabrics, as two colors may be placed side by side without bleeding. They have very high toxicity; some colors are mutagens, carcinogens, teratogens, and tumorigens.

# Synthetic Colorants

Synthetic colorants are manufactured and cost less than natural colorants, are offered in a diverse range of colors, are more colorfast, and are easy to apply.

Chromophores are an essential part of the colorant's chemical structure and are partly responsible for a chemical's ability to project color. Chromophores are limited in terms of the fibers upon which they can be used (i.e., limited to certain dye classes), the number of hues possible, the intensity of color required, and/or the cost.

More than 50 percent of all commercial colorants contain one or more functional chromophores known to the azo group. Many dye classes make use of azo groups (e.g., direct, azoic, reactive, acid, basic), as do some pigments, so their presence is not limited to any particular textile fiber or substrate. Under certain conditions, they can break down to form aromatic amines, which can then be released from the fabric or garment and may be carcinogenic.[3] The use of certain colorants that contain azo groups—mostly those that can release higher concentrations of amines—is forbidden in many parts of the world. This includes around a dozen acid dyes (normally used with nylon and wool) and numerous direct dyes (used with cellulosic fibers).

Anthraquinone is another common chromophore widely used for vat dyes.[4] Anthraquinone chromophores are found in vat, reactive, disperse, and acid dyes and in some pigments, so they can be used on cellulosic fibers such as cotton or rayon and on synthetic fibers such as polyester and nylon. Anthraquinone colorants are often brighter than their azo counterparts but are limited in terms of shade depth.

Disperse dyes are water **insoluble** and typically used for synthetic fabrics. Disperse dyes require a dispersing agent in the dye solution, which usually comes in the form of a surfactant that stabilizes the dye. A **surfactant** is a substance that lowers the surface tension between two liquids or a liquid and a solid (e.g., emulsifiers, detergents, dispersants). Surfactants typically end up in water discharge and contribute to high biological oxygen demand and chemical oxygen demand.[5] BOD refers to the milligrams of dissolved oxygen needed by aerobic biological organisms to break down organic material in one liter of water during five days at 208°C. It is often used as a measure of the degree of organic pollution of water. **Chemical oxygen demand (COD)** is a test used to indirectly measure the amount of organic compounds in water. It is typically used to measure the amount of organic pollutants in surface water or wastewater.

### Potential Impacts

Chemicals used in textile and garment dyeing and printing are developed to be resistant to environmental influences. This durability sometimes limits the biodegradability of colorants and makes them difficult to remove from wastewater generated by dyeing or printing processes.[6] Conventional treatments tend to transfer waste from one place to another. For example, solids extracted from wastewater are sometimes hazardous and are disposed of in special landfills, where they can cause groundwater contamination, gas formation, and noxious odors.

Certain types of dyes are suspected carcinogens[7] and mutagens,[8] while other disperse dyes are known to have a sensitizing effect on skin and should be avoided. Turquoise blue and greens contain metals, such as copper and nickel, as part of the dye molecule. Metals can cause toxicity in aquatic environments. Metal-containing colorants can be replaced with colorants that do not contain metals or contain lower metal content, though this is sometimes at the expense of colorfastness.

## Natural Colorants

Natural colorants are produced or extracted from plants, arthropods and marine invertebrates (e.g., sea urchins and starfish), algae, bacteria, fungi, and minerals. Sources for natural colorants include cochineal, mollusks, roots, bark, lichen, leaves, flowers, and other natural matter.

In order to achieve acceptable colorfastness, mordants are almost always necessary to properly fix natural colorants to textile and apparel substrates. Mordants increase colorfastness by combining with both the colorant molecule and the fiber molecule. The most commonly used mordants for natural dyes are tannic acid, aluminum, iron, and some other metal salts. Historically heavy metal salts including chromium, copper, and tin were used in natural dyeing, though they are typically eschewed today because of their toxicity and the instability of the colors they produce.

A major challenge with natural colorants is producing natural colorants in large quantities, at a reasonable cost, and achieving comparable colorfastness.

Obtaining a full palette of colors using natural colorants remains a challenge, as does repeatability, lightfastness, and durability to wear. Furthermore, in production, for some natural dysteuffs it is extremely difficult to produce the same natural dye shade twice, even when using exactly the same dyeing technique and procedure. For these reasons, natural dyes remain niche in the commercial fashion world and are well suited to the "slow fashion" movement, which emphasizes locality, difference, and diversity, though new technologies are working to bring natural dyes to wider economic viability.

**BOX 23.2**
**AVOID THESE DYE COLORS**

These dyes are carcinogenic or mutagenic colorants and should be completely avoided: CI Basic Red 9, CI Disperse Blue 1, CI Acid Red 26, CI Basic Violet 14, CI Disperse Orange 11, CI Direct Black 38, CI Direct Blue 6, CI Direct Red 28, and CI Disperse Yellow 3.

**TABLE 23.1 Natural Colorants**

| Natural Colorant | Dye Color |
|---|---|
| Cochineal beetle | pink, red, fuschia, purple |
| Oak gall | brown, grey, black |
| Sunflower seeds | purple, grey, green, orange, ochre, blue (dependent on species of flower) |
| Madder root | red, orange, coral, pink |
| Indigo | blue |
| Weld | yellow |
| Dyer's Coreopsis | yellows and oranges |
| Cutch | browns and ochre |

## Environmental Impacts

Metal mordants contribute to aquatic toxicity, if left untreated in the wastewater. Chromium's toxicity varies depending upon which form it is in: chromium(VI) is a known human and environmental toxin and carcinogen. Chromium(III) and chrome metals are generally accepted as safe. Tin is generally thought to be safe for humans to ingest up to levels of 1,400 milligrams/liter. In its inorganic form (e.g., wiring, jewelry, pennies), copper is a neurotoxin and carcinogen if ingested. All of these metal salts are harmful to

**Figure 23.2**
Naturally dyed wool yarn dyed with natural plant pigments.
*Peter Anderson/Dorling Kindersley/Getty Images*

human health if aspirated as a powder, which is the form most mordant metal salts take. Success has been achieved in some experimentation with dyeing in unlined copper pots (other metals may be used) as a way to take advantage of the metal's color-modifying properties without requiring its use in a more dangerous powdered form, but further research and development is needed.

# Printing: Application of Color

Printing is the patterned application of color. Since the colorants and **auxiliary chemicals** used in printing are similar to those used in dyeing, many of the environmental concerns are shared between these processes. However, to obtain the sharply defined, precise, reproducible patterns typical in printed textiles it's necessary to use special liquids, known as pastes or inks, that have a high degree of viscosity (i.e., they're in a "gel" state). The printing ink, which typically includes colorants, binders, softeners, thickeners, and other auxiliary chemicals, can be directly applied to the substrate using mesh screens or engraved rollers or ink-jet printers, or can be indirectly applied to the substrate using pre-printed transfer paper.

The two main types of printing inks are pigmented emulsions and inks.

## Emulsion Inks

**Emulsion inks** are used mainly for direct printing of fabrics and are typically water based, comprising a binder and cross-linking agent. A **cross-linking agent** is a chemical substance used to link polymer chains together. Emulsion inks can be solvent based, although their use is rare in textile and garment printing.

**Table 23.2 Environmental Impacts of Printing Inks**

| Type of Ink | Impacts |
|---|---|
| Emulsion inks | • Solvent-based inks have high volatile organic content (aliphatic, aromatic, and oxygenated organic solvents) and can cause problems with air pollution and waste disposal. **Aliphatic solvents** (e.g., hexane, gasoline, kerosene) are composed of open-chain (noncyclic) hydrocarbon chains, are often highly flammable, and some are neurotoxins. **Aromatic solvents** (e.g., benzene, toluene, xylene) are composed of ring-formation (cyclic) hydrocarbons and are often carcinogenic and highly flammable. **Oxygenated solvents** (e.g., alcohols, glycol ethers, methyl and ethyl acetates, ketones) are composed of molecules that contain oxygen. |
| **Plastisol inks** | • Plastisols do not biodegrade. One by-product of their production (and of their disposal via incineration) is dioxin, an acutely toxic substance.<br>• If landfilled, heavy metals sometimes used as PVC stabilizers, such as lead or cadmium, can contaminate groundwater.<br>• Plasticizers are often phthalate esters, which may leach out of the print or may evaporate and be released during drying, either during the production process or in the home. Exposure to phthalates is known to cause adverse health effects, and several phthalates are banned by California's Prop 65 law. |

**Figure 23.3**
Industrial screen printing with plastisol inks.
*John Spannos/iStock*

# Plastisol Inks

Plastisol inks are primarily used for direct and indirect (transfer) printing of garments and are typically vinyl resin (**PVC**) based. **Direct printing** describes printing techniques where the pigment or dye is applied directly to the textile. **Indirect printing** (or transfer printing) refers to a process by which the colorant is applied first to a carrier material and then transferred to the textile via the application of heat and/or pressure. In garment printing, PVC serves as a binder that melts into, or fuses with, the garment while bearing the solid pigment. Plastisol inks contain plasticizers, which soften the naturally rigid PVC to give it the flexibility to keep from cracking. When used as a transfer, the plastisol ink is screen printed onto a release paper and cured to a dry film, which is then stored until being transferred onto a garment using a heat transfer process.

# Printing: Discharge (Removal of Color)

Whereas printing is the patterned application of color, discharge printing is the patterned removal of color. In other words, the fabric or garment is dyed prior to printing (commonly known as the "ground" color or shade) and then printed with a paste or ink containing a chemical discharge agent. The discharge agent is capable of destroying the **chromophoric** system of the original colorant(s) under appropriate conditions, thereby severely degrading/fading the color or removing it altogether. Many reducing agents, oxidizing agents, acids, salts, and alkalis can function as discharge agents. All colorants react differently to these agents; some are dramatically affected, while others are

**Table 23.3 Environmental Impact of Discharge Printing**

| Type of Ink | Impacts and Considerations |
|---|---|
| Zinc formaldehyde-sulphoxylate | Releases formaldehyde during the discharge reaction process. Formaldehyde is retained in the fabric. Formaldehyde is a toxic air pollutant and a volatile organic compound; it is allergenic and/or carcinogenic in certain conditions and is heavily regulated. |
| Thiourea dioxide | Requires steaming and thorough washing after the discharge print paste is applied. |

largely or completely resistant. If more-resistant and less-resistant colorants are combined into one dyed ground shade and then discharge printed, the area to which the agent is applied creates a shift in hue. Colorants that are resistant to the discharge agent can also be included in the print paste itself so that they effectively "replace" the discharged colorant(s). This replacement color is sometimes referred to as the "effect" color.

The most common reducing agents used in discharge printing are metal salts of formaldehyde-sulphoxylic acid, such as zinc, sodium, or calcium formaldehyde-sulphoxylate. An alternative, but less frequently used, agent is thiourea dioxide.

### Zinc Formaldehyde-Sulphoxylate

Zinc formaldehyde-sulphoxylate (ZFS) is particularly popular because it helps to cure the acrylic binders commonly included in the discharge print paste. When combined with appropriate inks and a humectant, ZFS can be used in dry as well as wet or moist heat conditions. A **humectant** is a hygroscopic substance used to maintain moisture.

### Thiourea Dioxide

Unlike formaldehyde-sulphoxylates, thiourea dioxide neither contains nor releases formaldehyde. However, its effectiveness in discharging color is limited to a narrower range of colorants, and its effect is noticeably weaker than formaldehyde-sulphoxylates. Thiourea dioxide also requires steaming and thorough washing after the discharge print paste is applied.

**Figure 23.4**
Example of discharge printing.
*Kim Irwin/Fairchild Books*

## Dyeing: Reactive and Direct Dyes

When reactive or direct dyes are used to dye cotton, the use of salt—usually sodium chloride or sodium sulfate (also known as Glauber's salt)—is necessary to promote exhaustion (uptake and fastness) of the dye onto the cotton substrate. Sodium chloride is less expensive and contains more sodium per unit mass than Glauber's salt, so less salt is needed in the dye solution. However, Glauber's salt is less corrosive (particularly to dyeing machines) and produces brighter shades when used with some classes of dyes. In general, reactive dyes require five to ten times more salt than direct dyes, and dark shades require five to ten times more salt than light shades.

In addition to dye class and shade depth considerations, the amount of salt required in dyeing cloth is dependent upon the volume of the dyebath solution in relation to the mass of the material being dyed. This "liquor ratio" can vary widely depending on the dyeing equipment used. Some garment dyeing machines require as much as 400 gallons of dyebath for every pound of material dyed, whereas the commonly used jet dyeing systems require eighty gallons or less. In other words, the former would require five times more salt in the dyeing process than the latter.

### Environmental Impacts

Salt use in textile dyeing is a serious environmental issue because it is used in such large quantities and is therefore a major source of aquatic toxicity in wastewater.[9] Moreover, the removal of salt from wastewater is extremely difficult and expensive using current treatment methods.

Low- or no-salt developments are therefore of great interest in the chemical dye industry. Direct dyes use less salt than reactive dyes, but their attraction to cotton is relatively weak so they are often treated with special fixing agents to improve colorfastness. Cationic fixing agents (usually quaternary ammonium compounds) are commonly used to improve washfastness, and copper sulfate is sometimes used to improve lightfastness. Both are major concerns in terms of aquatic toxicity. **Cationic fixing agents** are a class of dye fixing agents composed of positively charged ions. They strengthen ionic bonds, which bond the dye to the fibers. Their health effects are still under investigation, but they can be toxic to animal, plant, and human life and may damage reproductive health.

## How to Reduce the Environmenal Impacts of Dyeing and Printing

### Naturally Colored Cotton

Naturally colored cotton has existed for thousands of years.[10] Naturally colored cotton has pigmentation in the center, or lumen, of the fiber, and the color depth and shade varies with growing conditions, location, and climatic factors. Generally speaking, natural colors range from shades of cream and tan to tones of brown, red, and green, although purple, mauve, gray, and black cottons are theoretically also possible.

Over the past twenty years, considerable work has been done to cross-breed colored cotton fibers to both expand the range of available colors and improve the fiber length and quality. As a rule, naturally colored cotton suffers from lower agricultural yields,[11] making it economically challenging to produce and therefore expensive to the mills and manufacturers. Brown fiber is typically two to three times the cost of white cotton fibers, and green fiber is approximately four times the cost of white fibers. Colored cotton fiber also lacks important quality characteristics such as fineness, length, and strength[12] and requires special handling through ginning and spinning, since the colored fibers can catch on equipment and contaminate the batches of white cotton fiber most commonly processed through the facility.

For these reasons, naturally colored cotton fibers are usually blended with white cotton to improve quality, facilitate processing, and reduce costs.[13] Though blending reduces the color intensity of the end product, washing the yarn or laundering the product in alkaline solution can enrich hues. The strength of 100 percent colored cotton can also be improved by plying several ends of yarns together (two- or three-ply), though this increases cost. Colored cotton yarns can also be plied with white cotton yarns to bring cost down. Colored cotton fiber properties can differ profoundly by color and are largely directed by the cultivation practices of the type of seed used. In terms of colorfastness to light, brown, and beige cottons generally outperform greens.

### "Right-First-Time" Dyeing

Without question, the most effective pollution prevention practice in coloring cloth is "right-first-time" dyeing. Corrective measures such as reworks, re-dyes, stripping, shade adjustments, top-ups, or "adds" are all chemically intensive and contribute significantly to pollution, since each corrective action increases colorant and/or chemical and water use. To improve "right-first-time" dyeing, use sampling and swatching in small batches with minimal inputs on the fabric to get the right shade before an entire run is dyed.

### Auxiliary Chemicals

Auxiliary chemicals can be selected to minimize or reduce the environmental impact of dyeing and printing processes. For example, acetic acid, which has a relatively high biological oxygen demand, is used in a variety of textile and garment processes for pH adjustment. Formic acid, which has a much lower BOD, or dilute mineral acids, which have no BOD, can sometimes be substituted for acetic acid.

In addition to chemical substitution, harmful auxiliary chemicals can sometimes be completely avoided by changing operating conditions. For example, "carriers" are organic chemicals that are often used as dyeing assistants when dyeing hydrophobic synthetic fibers such as polyester. In essence, carriers "open" the synthetic fiber, thereby increasing the rate of dyeing. The most common carriers are chlorinated benzenes and biphenyl. As a rule, carriers are extremely volatile and toxic, and they contribute significantly to toxic air pollution. (In chemistry, **volatility** refers to the tendency of a substance to vaporize.) But if dyeing takes place at a high enough temperature, carriers are unnecessary. In order to reach the required temperature (at least 129°C), a pressurized dyeing vessel is necessary. Many dyeing facilities have pressurized dyeing machines, but not all.

### Low-Liquor-Ratio Dyeing

A typical dyebath comprises salt, acids and alkalis, lubricants, and dispersing agents, all of which can contribute to pollution. These chemicals are measured in proportion to the volume of water used, so lower volume dyebaths greatly reduce chemical use and disposal in wastewater. Lower volume dyebaths also require less energy for heating. Standard liquor ratios range from 10:1 to 15:1 for many exhaust dyeing operations. Low-liquor-ratio machines are capable of dyeing at liquor ratios closer to 5:1, with some as low as 3:1. Some dye systems can operate effectively at room temperature, eliminating the need for heating altogether. It's important to note that low-liquor-ratio dyeing often limits the choice of dye class used (i.e., to more water-soluble dyes) and that existing equipment cannot normally be "retrofitted" to make it operate at lower liquor ratios; investment in new equipment is usually necessary.

### Dyebath Reuse

Dyebath reuse is the process by which exhausted hot dyebaths are analyzed for residual colorant concentrations, replenished, and—rather than being dispelled as wastewater—reused to dye further batches of fabric.[14] Dyebath reuse requires that colorants undergo minimal change during the dyeing process. Direct, disperse, acid, or basic dyes are therefore best for reuse applications. Dyebath reuse carries a greater risk of shade variation because impurities can build up and decrease the reliability of the process over the longer term. Capital is also required to purchase and install the appropriate infrastructure (e.g., holding tanks, pumps) to effectively reuse dyebaths. When properly controlled, some dyebaths can be reused for five to twenty-five cycles.

### Pad-Batch Dyeing

Pad-batch dyeing is a cold dyeing method mainly used for dyeing cellulose fibers (100 percent cotton and polyester/cotton blends) and can result in significant reductions in pollution and water and energy consumption (50–80 percent water and energy savings are common). No salt or chemical auxiliaries are necessary, and the colorant exhaustion is much higher (which means less color released into the wastewater). Moreover, quality is often more consistent compared to other exhaust dyeing techniques. Capital outlay is also low. However, pad-batch dyeing requires significant floor space to store dyed batches for long periods of time to allow the color to permeate the cloth. Many dye houses lack the needed space, and brands don't always have the time to accommodate the longer production processing time.

### Ink-Jet Printing (Also Known as Digital Printing)

Ink-jet printing is arguably the cleanest printing technology. Ink-jet printing is a noncontact printing method that works much like an office printer—droplets of colorant are propelled toward a substrate and directed to a desired spot. Colorant types that work best with ink-jet printing include reactive, vat, sulfur, and naphthol dyes, although acid, basic, and disperse dyes or pigments can also be used in some cases. Ink-jet printing eliminates the need for many printing auxiliary chemicals (e.g., thickener); eliminates the need for screen, squeegee, and machine cleaning (which also dramatically reduces water consumption); and reduces waste generated from strike offs. A **strike off** is a large sample of printed fabric used to test production methods. A strike off is usually multiple yards and tests pattern registration, repeat, and color matching to the original design. Though ink-jet printing machines represent a capital investment and production speeds are still relatively slow compared to analogue printing, ink-jet can offer significant savings for short production runs.

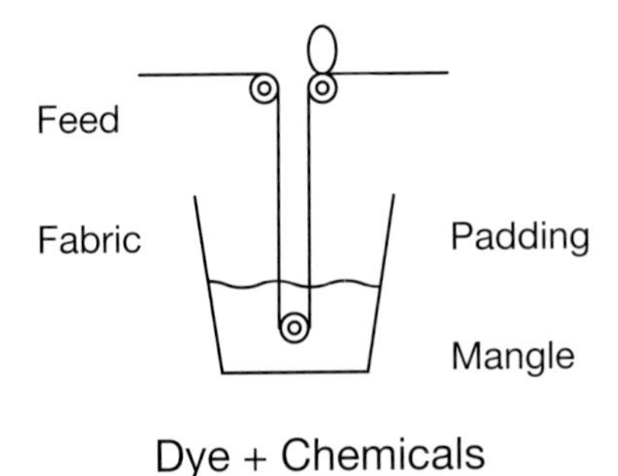

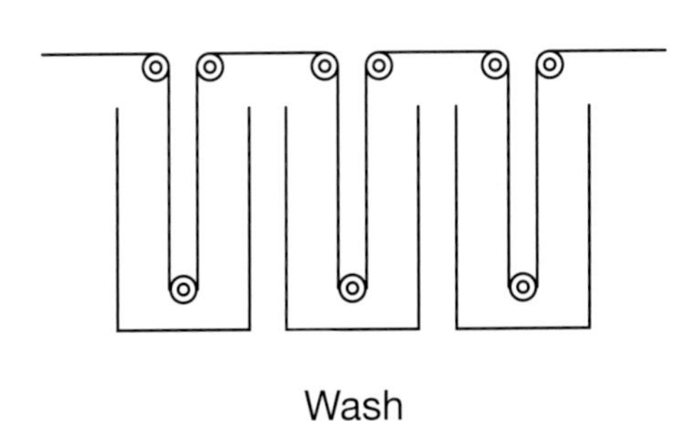


**Figure 23.5**
Pad-batch dyeing process.
*Yelena Safranova*

**Table 23.4 Optimize the Sustainability and Innovation Possibilities of Dyeing and Printing**

| Opportunity | Considerations |
|---|---|
| • Work with mills and vendors who use low-impact colorants (e.g., low COD/BOD, no metals, no formaldehyde).<br>• Forbid the use of carcinogenic/mutagenic colorants, restricted azo colorants, and organic solvent-based colorants. | • Carcinogenic, mutagenic, and teratogenic colorants are never acceptable. |
| • Work with printers who offer alternatives to PVC printing (e.g., resist printing, novel techniques such as REHANCE printing technique by TS Designs) and low- or no-formaldehyde, low- or no-metal plastisol variants. | • Hand feel, durability, and colorfastness will be improved with resist printing and techniques like REHANCE. |
| • Develop discharge prints with printers who use low-impact colorants and discharge agents.<br>• Avoid ZFS, heavy metals, and formaldehyde.<br>• Develop with printers who offer alternatives to discharge printing (e.g., resist printers, novel techniques such as REHANCE printing). | • May alter the product's aesthetic. |
| • Do not use mills and vendors who use carriers when dyeing synthetics.<br>• Promote mills that use high-pressure dyeing equipment. | • High-pressure dye systems generally use more energy. |
| • Promote lighter cotton shades and use of direct dyes in order to minimize salt volume and waste.<br>• Promote mills and vendors who use low- and/or no-salt dyeing techniques. | • If proper colorants are selected and more sophisticated equipment is used (e.g., automated dosing), salt can be reduced dramatically. |
| • Look for naturally colored cotton design and merchandising opportunities. Emphasize and feature positive physical and color performance properties. | • Colored cotton used in 100 percent produces weak yarn that can snap when making fabric. Use plied yarns in 100 percent or blend colored with white fiber to improve yarn strength. |
| • Find and promote mills and vendors who reuse dyebaths.<br>• Design tonal color range collections that are easier to work with in reuse dyebath systems. | |
| • Work with mills that pad-batch dye cotton to minimize salt, water, and energy use. | • The quality and price are excellent.<br>• Plan for longer manufacturing turnarounds. |
| • Work with ink-jet printers. Look for small-run niche opportunities with chase possibilities. | |
| • Promote proper wastewater treatment. | • Could vary based on geographic location of supplier. Consider working with suppliers who have verified through certification that water discharge is nontoxic. |
| • Promote the use of new technologies. | • Some newer technologies aim to eliminate water altogether and therefore also eliminate pollution associated with dyebath waste and more typical jet dyeing systems. These technologies, such as AirDye, can also create richer hues and achiever higher colorfastness than conventional systems.<br>• These technologies are most commonly used with polyester but could also be used with other synthetics. |

| Opportunity | Considerations |
|---|---|
| • Use natural colorants on protein fibers such as silk and wool, where their colorfastness and durability are optimized.<br>• Design natural dyes to fade beautifully. | • Natural colors are expensive and difficult to achieve lightfastness, washfastness, and long-term durability.<br>• Be prepared to accommodate different performance parameters (colorfastness, repeatability, etc.), or to design into the varied rates of fading, color to color. |
| • Know the source and cultivation details of natural dye plants and ensure they are sustainably grown.<br>• Promote mills and vendors who have in-depth knowledge of natural colorants and avoid heavy metal salt mordants. | • Just because you are using natural colorants does not automatically make it "good" or nontoxic. Make sure you are knowledgeable about all inputs throughout the process, including source of organic material for colorants.<br>• Also be aware of the volume of water used in your natural dye process and how that water is used post-dyeing. |
| • Use dyes that have been verified as safe from trusted industry-wide standards, such as Cradle to Cradle Certified™ and Bluesign approved. | • Must request that your suppliers use these dyes. Color palettes may be limited. |

### AirDye

Different technologies have been developed to manage the complex impacts of dyeing cloth. Heat transfer printing, for example, prints dye onto paper and then the paper carries the dye through a secondary process where the color is moved onto the fabric. Traditional heat transfer printing has poor fiber penetration, but new technologies can achieve greater penetration into the fiber by using dyes carefully developed/selected by "impregnating fabric fibers with color." AirDye, for example, can be printed with different colors or prints on the front and the back, and color is generally unlimited, though deep reds, violets, and fluorescents are currently not available. Besides reducing water and pollution associated with more typical jet dyeing systems, AirDye claims its process also creates richer hues and achieves higher colorfastness. These technologies are suitable only for synthetics and are used on polyester and nylon.[15]

## Potential Marketing Opportunities

- Simply state positive facts about the colorants used to dye the garments. (e.g., dyed or printed without using solvents, metals, salt, formaldehyde, etc.).
- Direct more costly technologies to higher-end fashion items (e.g., "Nature Dyed").
- Develop graphic symbols for more novel approaches such as recycled dyebaths and ink-jet printing to communicate with and inform the customer.
- Provide educational information online (e.g., social responsibility or brand website) detailing information about dyeing and printing in general and your company's commitments to switching to lower-impact dyeing processes.

# INNOVATION EXERCISES

## Design

1. Design a garment that emphasizes one or more of the sustainability benefits of dyeing or printing.
2. Design a garment that circumvents one or more of the environmental impacts of dyeing or printing.

## Merchandising

3. Create a fifty-word marketing communication message for consumers about the sustainability benefits of lower impact dyeing or printing processes.
4. Create a retail point-of-purchase sign or hangtag to promote fashion merchandise made from lower-impact dyeing or printing processes.
5. Go online and find three fashion retailers who sell fashion merchandise made with lower-impact dyeing or printing processes. What are the retailers and fashion merchandise offered? Why did they select this finish for their fashion merchandise? How do they promote the advantages of this finish to consumers? Do you believe their promotional strategies are accurate and effective? Why or why not?
6. Develop a marketing research protocol (e.g., focus group protocol, online survey, observational strategy) to determine consumers' perceptions of lower-impact dyeing or printing processes for fashion merchandise.

# REVIEW AND RECOMMENDATION

## Preferred/Not Preferred

While this chapter discusses an overview of the ecological impacts from dyes and printing, each dyestuff can have unique human ecological health characteristics. What makes it even more complicated is that each dye manufacturer can have varied formulations and added chemicals—information that is generally proprietary and not publically available. Leverage the techniques and alternative methods of dyeing discussed in this chapter.

Even more important, use dyes that have been verified as safe under strict standards, including Bluesign and Cradle to Cradle Certified™. You can request these dyes through your suppliers.

## Innovation Exercise Example

**Figure 23.6**
This design sketch uses fading as a desirable design aesthetic. In this example, natural dyes are used deliberately in areas of the garment to fade beautifully and in turn keeping the wearer engaged with its evolving look.

*Amy Williams*

JeanUrsula/iStock

# 24

# Finishing

Finishes used in the apparel industry are chemicals and treatments applied to fabric to enhance performance, durability, and longevity of wear. These treatments include soil and stain repellents, water repellents and waterproofing, antiwrinkle treatments, antimicrobials, flame retardants, and antistatic treatments.

Finishing applications such as water, stain, and odor repellents and flame retardants can greatly improve the performance of garments and textiles. While only a small amount of finishing chemicals are needed in the overall composition of a garment, these finishing applications can have significant bioaccumulative effects on people and the environment. **Bioaccumulation** refers to the accumulation of chemical substances in an organism at a rate faster than the organism is able to excrete or catabolize it. An example of a bioaccumulative substance is mercury, which has accumulated in sea life in places such as the San Francisco Bay Area, rendering these creatures unsafe for human and animal consumption.

# Benefits

### Water Repellents

Durable water repellents (DWRs) are a class of coatings added to fabrics at the factory. DWRs are applied to garments and products to allow for breathability and water repellency. Common factory-applied treatments are fluorochemicals.

### Stain Repellents

Stain-repellent finishes are used to provide stain, soil, and grease release and repellency to fabrics. Fluorochemicals are the most employed repellents used for textiles.

### Flame Retardants

Common flame retardants are brominated organic compounds; they are added to textiles and upholstery to delay the production of flames and prevent the spread of fire.

### Antimicrobials

Antimicrobials are used in applications such as socks, shoes, and activewear to prevent odor caused by the breakdown of sweat.

# Potential Impacts

### Water Repellents

Certain DWRs are known to have persistent bioaccumulative and toxicological effects on the environment due to the class of fluorochemicals used.

**Figure 24.1**
Durable water repellants are a class of coatings added to fabrics at the factory. DWRs are applied to garments and products to allow for breathability and water repellency.
*PeopleImages/Getty Images*

The durable water-repellent coatings used in the fashion and textile industry are currently not bio-based or biodegradable. Water-repellent coatings also inhibit recyclability of synthetic materials.

### Stain Repellents

The largest concern for chemicals used for soil- and stain-repellent finishes is perfluorooctanoic acid (PFOA), which is used in the manufacture of stain-repellent finishes for textiles. PFOA is also produced indirectly through the gradual breakdown of fluorochemicals.[1]

PFOA is very persistent in the environment and has been found at very low levels both in the environment and in the blood of the general U.S. population.[2,3]

Recycling of textiles with stain-repellent finishes is also very difficult. Fluorochemicals are currently being phased out by major industrial users.[4]

### Flame Retardants

Some flame retardants release hydrogen cyanide when set afire and can be deadlier than carbon monoxide.[5] Flame-retardant chemicals can be toxic, and some are suspected carcinogens.[6]

In the European Union (EU), the use of certain flame retardants is banned or restricted.[7]

### Antimicrobials

Organotins are often employed as antimicrobial treatments on textiles. The organotin compound tributyltin (TBT) is bioaccumulative, highly toxic to aquatic life, and persists in the environment. TBT is listed as a "priority hazardous substance" under European Union regulations and requires measures to be taken to eliminate its use.[8]

A recent report conducted by Greenpeace acknowledged organotins detected in several activewear products.

## Optimize Sustainability Impacts Through Alternatives for Finishing Treatments

### Water Repellents

- Investigate nonfluorochemical coatings, such as silicones, polyurethane (PU), and waxes. Although these coatings are recyclable on their own, they inhibit recyclability when applied to a different base material. Conversely, these coatings have the potential for recyclability if applied to a similar base material.[9]
- Work with manufacturers to create bio-based or biodegradable water-repellent finishes.
- Investigate recyclable waterproofing agents such as Sympatex. It is made of completely safe polyether/ester, a combination of polyester and polyether molecules is reportedly recyclable if applied to a similar base material (i.e., polyester). Sympatex contains zero fluorochemicals.[10]
- Investigate durable water repellents from renewable nontoxic resources, such as castor oil.

- Investigate PDF-free membranes for waterproofing.
- Investigate the use of fabrics that are not coated or laminated, but are instead densely woven to create a structurally waterproof textile, such as Ventile®.

### Stain Repellents

- Investigate short chain fluorocarbons that do not degrade into PFOA.
- Investigate stain-resistant finishes that do not involve the use of PFOA, such as finishes from DuPont.

### Flame Retardants

- Investigate nontoxic flame-retardant applications.
- Investigate the use of halogen-free flame retardants, such as those made by InnoSense LLC.
- Investigate using polyester as an alternative to textiles with a flame-retardant coating. Polyester is inherently flame retardant.

### Antimicrobials

- Investigate nontoxic biodegradable alternatives to organotins.
- UV curing of Chitosan as an antimicrobial finishing for textiles is still in development and could provide a bio-derived, nontoxic, biodegradable alternative to organotins.[11] **Chitosan** is a linear polysaccharide material made by treating shrimp and other crustacean shells with sodium hydroxide. Ongoing research has found that when it is cured with UV light, the substance shows potential as an antimicrobial surface treatment.
- Experiment with natural fibers that intrinsically repel odors, such as wool. Some natural dye plants, such as lavender, can also be effective as an antimicrobial treatment and can be applied without coloring effects.

**Figure 24.2**
Investigate the use of halogen-free flame retardants, such as those made by InnoSense LLC.
*Red Chopsticks/Getty Images*

## Availability

- Silicon-based water repellents are available globally. Royal DSM's Arnitel® VT, for example, contains no PDFs and is recyclable, breathable, and waterproof.[12]
- Alternative stain-repellent finishes are available from DuPont.
- Halogen-free flame retardants are available from some suppliers, such as InnoSense LLC.
- Work with suppliers that offer antimicrobials (Akzo Nobel, BASF, Arch Lonza, Dow, Troy Corp) to request antimicrobials that are nontoxic and safe for humans and the environment.

## Fashion Applications

- Water repellents are mainly used for outdoor clothing, such as raincoats and jackets.
- Flame retardants are typically found in children's clothing.
- Stain repellents are found in children's clothing and shoes, as well as both casual wear and activewear for adults.
- Antimicrobials are found in children's clothing, shoes, and socks, as well as activewear, shoes, and socks for adults.[13]

## Potential Marketing Opportunities

In 2014, Greenpeace produced a report titled "A Little Story about Monsters in Your Closet," that details the toxic effects of textile finishes and where the organization found them in children's clothing from well-known brands. Aside from this report, there is no significant consumer awareness around the toxic effects of textile finishes due in part to the fashion industry's primary focus on fibers and fabrics. As mentioned previously, while textile finishes can be less than 5 percent of the overall composition of a garment, they can still have a significant impact on the environment, especially if a large percentage of the company's business (in the performance gear markets, for example) is outerwear. It is important that, through your communications, you clearly state impacts associated with textile finishes used and explain in a positive way why you have chosen an alternative finish as well as how this affects your customers, your business, and the environment.

# INNOVATION EXERCISES

### Design

1. Research a finishing treatment that is a positive alternative to a commonly used finish and identify any aesthetic differences it might bring. Design a garment that utilizes that finishing technique.
2. Design a garment that circumvents one or more of the environmental impacts of one of the finishing techniques mentioned.

### Merchandising

3. Create a fifty-word marketing communication message for consumers about the sustainability benefits of lower-impact finishing processes.
4. Create a retail point-of-purchase sign or hangtag to promote fashion merchandise made from lower-impact finishing processes.
5. Go online and find three fashion retailers who sell fashion merchandise made with lower-impact finishing processes. What are the retailers and fashion merchandise offered? Why did they select this finish for their fashion merchandise? How do they promote the advantages of this finish to consumers? Do you believe their promotional strategies are accurate and effective? Why or why not?
6. Develop a marketing research protocol (e.g., focus group protocol, online survey, observational strategy) to determine consumers' perceptions of lower-impact finishing processes for fashion merchandise.

## REVIEW AND RECOMMENDATION

### Preferred

- Nonhalogenated soil and stain resistant additives/coatings
- Nonfluorochemical coatings, such as silicones, polyurethane, and waxes for waterproofing
- Textiles that do not require waterproofing, such as Ventile®
- Nontoxic biodegradable alternatives to organotins for antimicrobials
- UV curing of chitosan as an antimicrobial finishing for textiles
- Natural fibers that intrinsically repel odors, such as wool
- Investigating the use of halogen-free flame retardants, such as those from InnoSense LLC
- Investigating the use of polyester as an alternative to textiles with a flame-retardant coating since polyester is inherently flame retardant

## Innovation Exercise Example

**Design a garment that circumvents one or more of the environmental impacts of one of the finishing techniques mentioned.**

**Figure 24.3**
In this example, wool, a natural antimocrobial, is used in children's clothing to circumvent the need to use an antimocrobial treatment. Darker-colored yarns are used in areas where children are most likely to stain.
*Amy Williams*

cookelma/iStock

# 25

## Garment Washing

In addition to improving or softening the hand feel of products, garment washing affects the aesthetic of the product, often by imparting a "worn in" or "aged" appearance. Garment washing has become an indispensable tool for apparel designers to manipulate garment aesthetic and to impart unique decorative effects, particularly for denim. The umbrella terms "garment wet processing," "garment wet and dry processing," "garment finishing," or just "garment processing" can be used interchangeably to describe many different techniques, all designed to alter the garment's hand feel or aesthetic in some fashion. "Garment washing" generally entails treatments involving water and chemicals.

Dry procedures are used primarily for localized or even "patterned" abrasion effects and include techniques such as sandblasting, hand sanding, brushing, grinding, cutting (holes/patches), and more.

Wet garment washing processes involve the use of numerous chemicals, depending on the exact nature of the process. Most wet processes are designed to abrade, decolorize, and/or soften the garments. Although the techniques are generally intended as an all-over treatment, the degree of abrasion, decolorization, and/or softening can and does vary significantly within and between garments in a typical load. For example, thick and/or more-exposed areas of the garment (such as hems/seams) absorb more of the mechanical or kinetic energy during tumbling and may therefore be more abraded and/or decolorized (faded) than flat areas. Conversely, tightly constructed areas of the garments may end up less decolorized than less-dense areas since their ability to absorb chemicals (e.g., bleach) may be hindered.

By far, the most involved and intensive wet treatments are applied to denim products, although many of these same treatments are now applied to other woven bottoms, woven tops, and even knit garments.

Two basic types of equipment are used for garment washing: 1) side-loading horizontal washers (commonly referred to as belly washers) and 2) front-loading rotary washer/extractors. There are numerous variants of each machine type.

Rotary washer/extractors, the more expensive of the two, generally provide many more options to control/optimize wet treatments, including advanced liquor ratio (water-to-fabric ratio) control, heating, and colorant/chemical add systems. As a rule, they provide more opportunities for waste minimization than belly washers.

After garment washing, large open-pocket tumble dryers are typically used to dry apparel. Smaller units may be heated electrically, while larger units are typically steam or gas heated. Modern tumble dryers have relatively sophisticated controls (e.g., moisture sensors) that help to minimize energy use.

# Steps in a Typical Denim Garment Washing Process

### Desizing/Scouring

Woven denim products must be desized before further garment washing since they still contain sizing agents applied to the warp yarns. The most common sizing agent is starch, which is typically removed by adding amylase enzymes to break down the starch molecules into water-soluble sugars, making them easier to remove.

Another common sizing agent is polyvinyl alcohol (PVA). PVA is relatively water soluble, provided it is the right "grade." No enzymes or oxidizing agents are necessary to remove water-soluble sizing agents; they can be rinsed from the garments simply by using adequate washing temperatures and times with a good detergent. PVA can create a high chemical load in the wastewater and should be reclaimed for reuse.

In addition to sizing agents, other "top finishes" are sometimes applied to fabrics for purposes of lubrication (e.g., sewing lubricants, **sanforizing lubricants**). Garment scouring removes these top finishes by rinsing with an appropriate detergent. Light scouring and desizing softens denim garments drastically.

### Wet Abrasion

Wet abrasion techniques are used to create a natural-looking (uneven) worn and faded effect, ranging from slightly to very uneven. Wet abrasion increases seam contrast since thicker regions of the garment tend to abrade more readily than flatter regions. In its most basic form wet abrasion entails tumbling wet garments in the presence of pumice stones (or an appropriate substitute). Commonly known as "stone washing," this technique can create a wide array of effects by adjusting the amount of water or number of stones, the size or shape of the stones, the tumbling time, and the mass ratio of stones to garments.

Sometimes, stones are presoaked in an oxidative chemical solution (i.e., bleach) prior to tumbling with the garments. This increases the decolorizing potential of the stones through the release of bleach to specific areas of the fabric as the garment and the stones collide. Common oxidizing agents used for this purpose include chlorine derivatives (e.g., sodium or calcium hypochlorite) and potassium permanganate. In fact, the once-popular "acid wash" was achieved by tumbling garments with stones that had been presoaked in potassium permanganate.

- Visit **Chapter 22: Bleaching** for information on the environmental impacts of bleaching.

Cellulase enzymes can also be used to accelerate wet abrasion effects (by removing or weakening the surface fiber) and can reduce, or in some cases eliminate, the need for stones altogether.

Wet abrasion is usually followed by a quick rinse intended to remove any remaining loose dyestuff and/or residual dust from stones or other abrasive materials and to deactivate remaining cellulase enzymes if necessary.

### Bleaching

Bleaching is often used to lighten the color of garments overall, to brighten the indigo dye used on denim products, and to remove indigo dyestuff that may have deposited on the (undyed) filling yarns during the wet abrasion process.

Most garment bleaching is done with chlorine derivatives (usually sodium or calcium hypochlorite). Hypochlorite is a strong bleach that is reactive enough to work well at cooler temperatures and is effective at removing certain dyestuffs from garments.

- Visit **Chapter 22: Bleaching** for information on the environmental impacts of bleaching.

### Brightening

Sometimes referred to as "top brightening," this garment washing technique may be utilized after bleaching to further whiten, or brighten, the decolorized areas of the garment, thereby enhancing the contrast between the light (or white) and dark areas in the fabric. This is sometimes accomplished using a milder bleaching agent (e.g., hydrogen peroxide) or by using an optical brightener. Optical brighteners (also known as fluorescent whitening agents, or FWAs) are colorless dyestuffs that have the ability to absorb invisible UV radiation and retransmit it as visible (white) light.

### Tinting/Overdyeing

The application of additional colorant to garments that have already been dyed and/or printed is known as tinting or overdyeing. If any areas of the garment are white (e.g., filling/weft yarns in denim garments), these will fully absorb the colorant, but dyed or printed areas will also pick up a degree of color. Any type of colorants (e.g., dyes, pigments, metal salts) may be used to tint fabrics, depending on the substrate. Overdyeing with one color changes the hue of the preexisting colors and tends to "unify" the look of the print, often imparting a more vintage or dusty appearance to the fabric. For example, overdyeing with a blue shade will turn browns into warm, deep grays and will turn grays into soft blues. Overdyeing with a red shade will turn browns into deep rust and will turn grays into soft reds. Overdyeing with secondary shades can result in even softer and more complex effects.

- Visit **Chapter 23: Dyeing and Printing** for more information on impacts associated with dyeing.

### Softening

The final step of most garment washing operations is softener application, which can enhance the garment's hand, drape, abrasion resistance, and even tear strength. There are many different types of chemicals that can function as softeners, including sulfates and sulfonates, amines and quaternary amines, ethylene oxide derivatives, and hydrocarbon waxes. Softener selection is primarily a function of the desired hand feel: dry (petrochemical/polyethylene), greasy (organic/fatty derivatives), or slick (silicone). Roughly one-third of the chemicals used as softeners are silicone based. Softeners work by reducing the coefficient of friction of fibers and yarns.

# Potential Impacts

Although dry processing techniques involve no chemicals, they do create environmental impacts, including extraction of abrasive media from natural habitats, the transport of material to the processing facility (often surprisingly long distances), and the landfilling of spent abrasive media. Dry techniques such as sandblasting can also involve considerable occupational health and safety hazards for operators, and proper safety precautions, such as appropriate personal protective equipment and adequate ventilation, must be in place—but aren't always taken.[1] Although sandblasting has been banned for decades in Europe, the practice is still abundant in China, India, Bangladesh, Pakistan, and parts of Northern Africa—countries with a high likelihood of poor enforcement of proper safety precautions.[2,3]

Garment washing is a relatively water-intensive process, and may also be energy and chemical intensive, depending on the nature of the wash used. The environmental impacts of garment washing include the discharge of chemicals (surfactants, chelating agents, acids, alkalis, oxidizing agents, reducing agents, heavy metals, etc.) and colorants into water systems, which contribute to aquatic toxicity and/or high biological demand (BOD) or chemical oxygen demand (COD). High BOD and COD create environments that are hostile to aquatic plants and animals and may create problems with water reuse. Any color removed from the garments during the garment washing process is also dispelled to the wastewater and may create problems with photosynthesis for aquatic plant life.

In order to promote the permanence of color on a textile substrate or garment, colorants and other chemicals used in textile and garment dyeing and printing are developed to be resistant to environmental influences. This durability sometimes limits the biodegradability of colorants and makes them difficult to remove from wastewater generated by dyeing or printing processes.[4]

In terms of its environmental impact, hypochlorite used in bleaching breaks down into table salt, oxygen, and water. But if hypochlorite is exposed to organic material before it breaks down, it can react with that material to form halogenated organic compounds (**organochlorines**). Halogenated organic compounds are persistent, toxic compounds that may bioaccumulate in the food chain, are known teratogens/mutagens and suspected human carcinogens,[5] and may cause reproductive harm. Hypochlorite can also react with acids, ammonia, and even dirt particles to form toxic gases.

Toxicity and biodegradability of chemicals used as softeners are primary considerations. As a rule, fatty derivatives are highly biodegradable, whereas petrochemicals are not. Silicone is highly resistant to biodegradation by microorganisms (such as those used in biological wastewater treatment) but will degrade in soil (e.g., in a landfill).

Impacts from colorants for tinting/overdyeing include toxicity to air, water, and skin.

- Visit **Chapter 23: Dyeing and Printing** for more information on impacts associated with dyeing.

# Techniques to Minimize Pollutants, Water Use, and Energy Consumption

### Fabric Selection

One of the keys to reducing the environmental impact of garment washing processes is to select fabrics with desired garment hand feel and aesthetic qualities engineered into the construction. If fabrics are physically engineered to exhibit desired qualities, the intensity (and by extension, the environmental impact) of many garment washing treatments can be minimized.

For example, the hand feel of garments can be dramatically altered by modifying fiber diameter and cross-section, fiber length, fiber tenacity and modulus, yarn twist, yarn count, yarn hairiness, fabric stitch density, etc., thereby minimizing the need for hand-feel modification in garment washing. In many cases, only slight physical modifications are necessary (i.e., they are visually undetectable).

Similarly, the impacts of garment washing can be reduced significantly by selecting colors close to the desired final hue. Fifty percent or more of the colorant for deep shades is removed via abrasion or bleaching in garment washing. Selecting a color closer to the desired garment shade after wash reduces the degree of decolorization (and associated energy, dyestuff, and waste) necessary.

### Water Reuse

In addition to minimizing the amount of water coming into a textile mill, water conservation can also occur after the wet processing is complete. A typical garment washing process may involve several wash cycles (e.g., desizing, wet abrasion, bleaching) as well as an assortment of rinses between cycles. This requires the garment washing machine to be drained and refilled numerous times. It's not unusual, for example, to use thirty-five or more gallons of water per garment during the garment washing process. In order for this water to be recycled and/or reused, it must contain little or no chlorine and have low metal content and low salt concentration (e.g., chloride and sulfate). Residual dyes and pH level are also of concern.

Some garment washers have reduced water consumption by 50 percent or more by reusing process water. Some municipalities have even started marketing recycled water (e.g., water treated via reverse osmosis) to industrial customers. In fact, treated and recycled water is sometimes more consistent in terms of its impurities than potable water.

### Frequency of Machine Cleanings

Total water consumption in garment washing is also affected by the frequency of machine cleaning. In general, scheduling machines to process progressively darker shades—from light to dark—minimizes the need to clean the machine between each color batch.

### Low-Liquor-Ratio Washing

One of the most important considerations, from a water and energy consumption perspective, is liquor ratio. Liquor ratio is the weight of the chemical bath (including the water) divided by the weight of the material (garments) being processed. Garment washing machines are available in a variety of sizes, and loads can vary widely depending on the nature and scale of the order. If the load size is small and a large garment washing machine is used, the liquor ratio, the water volume, and the energy used to heat that water will all be higher than necessary. Liquor ratio also affects the speed and level of fabric abrasion. In higher-liquor-ratio machines, garments and abrasive materials come into contact with each other less than in low-liquor-ratio circumstances. High liquor ratios therefore require more time (and energy) to achieve similar abrasion levels than low liquor ratios. As a rule, front-loading rotary washer/extractors have more flexible controls to accommodate various load sizes, enabling optimal water and energy use and minimizing waste.

### Proper Chemical Selection

Another important element of pollution prevention is chemical selection. A wide variety of surfactants, chelating agents, oxidizing agents, reducing agents, enzymes, lubricants, colorants, and other chemical types are routinely used in garment washing. Vendors generally choose chemicals based on their performance characteristics (effectiveness) and price but must also factor environmental considerations such as toxicity, BOD, and COD into chemical selection decisions. In addition, the biodegradability of each chemical is of prime importance. For example, alkyl phenol ethoxylates (APEOs), a common class of surfactant used in garment washing, are undesirable due to their poor biodegradability, their toxicity (including that of their phenolic metabolites), and their potential to act as endocrine disrupters. APEOs are banned in Europe, and there are a host of wetter/scour alternatives readily available. Overseas garment washing operations may still use APEO surfactants because of their low cost and good performance characteristics. In general, chemicals and their processes should be selected to be the most benign. For some processes, enzymes can replace chemicals and include amylases used for desizing, cellulases used for wet abrasion, and laccases used for bleaching.

### Lasers

Another alternative to traditional processes of decolorizing fabrics is the use of lasers.[6] Depending on its wavelength, the laser can either 1) be absorbed by and decompose the colorant, or 2) be absorbed by and alter the surface chemistry of the fabric. The latter technique in particular has great potential to replace traditional wet abrasion and mechanical abrasion techniques such as hand sanding because it closely emulates the results these traditional abrasion processes achieve. Lasers for this purpose are already commercially available, and several are in use in laundries around the world.

### Bleaching Alternatives

There are two technologies designed to replace chlorine derivative bleaches: ozone- and enzyme-based processes. Ozone can be used with no water at all and is very effective at fading pre-dyed/preprinted fabrics and garments. Laccase is an enzyme that has proven effective at decolorizing or fading pre-dyed/preprinted apparel products. Enzymes are biodegradable (no-waste products) and typically work well at low temperatures, thereby minimizing energy consumption.

- Visit **Chapter 22: Bleaching** for more information.

### Combination/Elimination of Garment Washing Processes

Another possibility for pollution prevention is the combination, or even the elimination, of specific garment washing processes. For example, denim desizing and wet abrasion have long been performed using two completely separate garment washing treatments, each with its own environmental impact. Desizing is performed using an amylase enzyme or oxidizing agent, followed by a wet abrasion treatment using stones, cellulase enzymes, or both. It is sometimes possible to combine these two cycles into one, which significantly reduces process time as well as water, energy, and chemical consumption. However, combining desizing and wet abrasion processes can present specific technical issues, such as severe streaking and back staining from the large amounts of dye present in the desizing bath.

**Figure 25.1**
Denim lasering at Saitex in Vietnam, committed to providing a safe planet for our future generations.
*Saitex*

**Figure 25.2**
Ozone machine at Saitex in Vietnam.
*Saitex*

These issues can be overcome by using two specific types of cellulase enzymes in combination—one designed for abrasion assistance, and the other designed for streak reduction/prevention. Enzymes are added to the processing bath in a specific sequence or are selected to have an appropriate "dormant" period (i.e., the enzyme is not activated until the proper time during the washing cycle). This strategy is sometimes called a "combi-process."

## Waste Minimization/Source Reduction

Strong consideration should be given to whether a garment washing process is truly warranted. On certain products (e.g., denim), some form of garment washing is often necessary—to remove the sizing agents present on the warp yarns, for example. On other products, such as knit tops, garment washing may not always be required. For example, where garment washing is performed to reduce the hairiness/pilling propensity of knit garments, fiber selection (e.g., less short fiber or lower-tenacity fiber) and/or modifications to the yarn (less twist) can sometimes eliminate that need. Where garment washing is performed to reduce the torque/skew or shrinkage in garments, some procedures in the manufacture of the fabric, such as sanforizing, may suffice. And where garment washing is used to achieve a faded aesthetic, starting with a fabric shade closer to the desired garment shade after wash can significantly reduce the degree of decolorization (and associated energy, dyestuff, and waste) necessary.

## Eco-Aging

**Eco-aging** is an alternative to sandblasting created by Fimatex in Italy. It's a fading process that uses a vegetable mix composed of waste from food and is said to be biodegradable.[7]

**Table 25.1 Optimize the Sustainability Possibilities of Garment Washing**

| Design Opportunity | Considerations |
|---|---|
| • Look for opportunities to avoid garment washing. | • Fabric suppliers will need prompting to show physically engineered fabrics if these wash-saving processes have not been requested before. |
| • Select fabrics that are closer in shade to the desired garment shade after wash. | • Speaking with a technical person, rather than a salesperson, may be necessary. |
| • Encourage water conservation with existing suppliers and/or seek new suppliers who use water conservation techniques.<br>• Create awareness that water and energy conservation is important and possible without sacrificing hand feel or aesthetic. | • Water use by garment laundries varies widely. |
| • Promote proper wastewater treatment. | • Water treatment by garment laundries varies widely. |
| • Leverage the aesthetic differences that low-impact garment washes offer. Turn the differences into positive stories. | • Will likely have limitations. Understand what these limitations are and design around them. |

# Potential Marketing Opportunities

- Water conservation (gallons of water saved/per item) when ozone bleaching is used.
- Nonchlorine bleached can be claimed if nonchlorine bleach has been used and you have used a sustainable alternative.
- Laser treatments replacing conventional wet abrasion finishes could appeal to an increasingly tech-savvy consumer base.
- Educational information provided on the social responsibility or brand website could detail information about lower-impact processes for washing and finishing and bleaching.

# INNOVATION EXERCISES

## Design

1. Design a garment that emphasizes one or more of the sustainability benefits of garment washing.

2. Design a garment that circumvents one or more of the environmental impacts of garment washing.
3. Research one alternative to traditional garment washing that has a sustainability benefit. Design a garment that emphasizes its unique aesthetic.

### Merchandising

4. Create a fifty-word marketing communication message for consumers about the sustainability benefits of alternatives to traditional washing processes.
5. Create a retail point-of-purchase sign or hangtag to promote fashion merchandise made from alternatives to traditional washing processes.
6. Go online and find three fashion retailers who sell fashion merchandise made with alternatives to traditional washing processes. What are the retailers and fashion merchandise offered? Why did they select this finish for their fashion merchandise? How do they promote the advantages of this finish to consumers? Do you believe their promotional strategies are accurate and effective? Why or why not?
7. Develop a marketing research protocol (e.g., focus group protocol, online survey, observational strategy) to determine consumers' perceptions of alternatives to traditional washing processes for fashion merchandise.

## REVIEW AND RECOMMENDATION

### Preferred

- Garment washing methods that use alternatives to bleaching
- Proper chemical and fabric selection
- Laser methods

### Recommended Improvements

- Proper wastewater treatment

## OTHER SOURCES

1. Dupont n.d.
2. Grose and Fletcher 2012

## Innovation Exercise Example

**Research one alternative to traditional garment washing that has a sustainability benefit. Design a garment that emphasizes its unique aesthetic. In this example, a laser is used to mimic traditional washing techniques. The laser can be absorbed by and alter the surface chemistry of the fabrics to replace traditional wet abrasion and mechanical abrasion techniques.**

**Figure 25.3**
In this example, a laser is used to produce effects on denim to replace traditional wet abrasion and mechanical abrasion techniques. The laser can be absorbed by and alter the surface chemistry of the fabrics.
*Amy Williams*

BahadirTanriover/iStock

# FUTURE FIBERS: PROCESSING

This Future Fibers part closer presents innovative alternatives to traditional ways of processing textiles. From innovative dye methods to pretreatments for textiles for ecologically improving dye practices to methods for washing, this is only a sampling of the many companies who are developing better ways of processing chemicals. These methods address only a piece of the overall impact areas for textiles. It is important that you as a designer are ensuring that you make educated decisions to address other impact areas of the fibers, including cultivation, animal welfare, and chemical processing, detailed in the previous chapters.

**Figure P3.1**
A factory and research laboratory where fabric is researched, produced, and manufactured in Taiwan.
*Justin Guariglia/age fotostock/ Getty Images*

**Part 3 Table 1 Innovative Alternative Methods for Processing Textiles**

| Company | Overview | Considerations | Availability | Applications |
|---|---|---|---|---|
| Applied Separations (www.appliedseparations.com)<br><br>*Elena_Rodalis/iStock* | • Applied Separations has developed technology and machinery used for a supercritical $CO_2$ waterless dye process. This method uses high-pressurized $CO_2$ gas and heat to impregnate technical textiles with dye color. After the dye is set into the fabric, the supercritical $CO_2$/dye mix is depressurized, the $CO_2$ changes to a gas, and all the spent dye falls out and can be reused. In production systems, the $CO_2$ is recycled, providing for a completely closed system and an entirely environmentally friendly approach to textile dyeing.<br>• Dyeing with supercritical $CO_2$ results in adulterant-free, dispersed dye compounds covering the entire color spectrum. The final dyed material does not need any further treatment, such as using water to remove unabsorbed dye.<br>• Energy output used in the Applied Separations' supercritical $CO_2$ waterless dye method uses less energy consumption than traditional dye methods. | • Cannot be used on natural fibers. | • Commercially available. | • Use with synthetic fibers, such as polyester. |
| ColorZen (colorzen.com)<br><br>*KaraGrubis/iStock* | • ColorZen is a pretreatment for cotton fibers that allows for less energy and less water use and requires no toxic chemicals for the dyeing of cotton. | • Ensure that cotton fiber used with ColorZen meets your criteria for ethically sourced. | • Commercially available. | • Use for cotton fiber applications. |

**Part 3 Table 1 Innovative Alternative Methods for Processing Textiles** *(continued)*

| Company | Overview | Considerations | Availability | Applications |
|---|---|---|---|---|
| Jeanologia (www.jeanologia.com/) <br> *bonetta/iStock* | • Laser- and ozone-finishing technology that saves energy, water, and chemicals. Eflow machine uses air bubbles to reduce the amount of water required to distribute product over the denim garments, therefore saving water and requiring less product because it is used more efficiently. Saves up to 95 percent water, 50 percent chemicals, and 75 percent energy. Jeanologia offers a washing machine adaptation that can be hooked up to existing machines. | • Could be higher cost than traditional washing. <br> • Ensure that fabric has been dyed sustainably. | • Commercially available. | • Use with knitwear, denim garments, and continuous fabric lengths. |
| Novozymes (novozymes.com) <br> *Leonid Andronov/iStock* | • Novozymes has developed enzymatic finishing solutions for processing textiles. These processes are applied in the use of enzymes for wet-processing biological nutrient textiles. The solutions can decrease dependence on chemicals, lower consumption of energy and water, and bring down costs. | • Ensure that the fibers you use are also addressing other potential impacts through the cultivation and processing of the feedstock and resulting fiber. | • Commercially available. | • Use with cotton knits and denim wovens. |
| DyStar® Indigo Vat dye, 40% solution (dystar.com) <br> *BahadirTanriover/iStock* | • This prereduced indigo liquid allows a cleaner denim production and reduces the use of sodium hydrosulfite usage by 60–70 percent, resulting in a cleaner water effluent. | • Requires testing in dyehouse before use. Potentially more costly than traditional synthetic indigo dyes. | • Commercially available. | • Use with denim wovens. |

# Optimize Sustainability Benefits

- Ensure that the fibers you use are also addressing other potential impacts through the cultivation and processing of the feedstock and resulting fiber.
- Ensure that finishes used don't prevent biodegradability or recyclability.

UIG via Getty Images

## Part 4

# RECYCLED/CIRCULAR TEXTILES

## Overview

Using recycled fiber is an intelligent solution to finding value in materials we would otherwise consider garbage or "waste." Using recycled fiber also achieves ecological benefits: 1) it slows the depletion of virgin natural resources, 2) it reduces textile waste building in landfills, and 3) it can ease the pressure industrialized farming places on the land to yield more virgin fiber.

It should always be your intention not only to use recycled textiles but to keep these textiles "circulating" within the system by using recycling technologies that can provide virgin-quality textile solutions. Upcycling is always the goal. While downcycling is an interim solution, this cannot keep fibers circulating in the system in perpetuity. The goal is always is to cut down on our reliance of virgin resources and to use what we have. This is intelligent materials use.

The following chapter focuses on types of recycling for textiles.

Brian Ach/Getty Images for H&M

# 26

# Recycled/Circular Textiles Technologies

There are two types of waste, or "input," that can be used for recycled fibers for apparel: **post-consumer waste** from used and discarded clothing products, and **post-industrial waste** from material collected during the product manufacturing stage.

Post-industrial (also known as pre-consumer) input utilizes material created during product manufacturing. Examples of post-industrial waste include selvage from weaving, fiber waste from spinning, cutting-room waste, and fabric remnants. Energy may be required to convert the waste into usable forms, and the waste may be used as a raw material in the textile plant or may be sold and used for some other purpose unrelated to textiles or apparel, such as stuffing or padding in automobiles, furniture, or mattresses, or as raw material for paper or for coarse yarns (e.g., yarn for mop heads, industrial belting, rope, or twine), and as insulation.

More than 50 percent of post-industrial textile waste is reused or recycled in some fashion.

Post-consumer input for recycling is generated from used and discarded clothing or other products, such as handbags, home linens, or footwear. These products are collected, deconstructed or disassembled, and either recycled and used as raw material for recycled textile fibers or sold and used for some other purpose unrelated to textiles and apparel. It would always be the intention to maximize a product's use stage as long as possible before recycling to capitalize on a garment's embodied energy and resources associated with its creation.

There are three main types of recycling for post-industrial or post-consumer apparel and textiles: mechanical, melt processing, and chemical.

## Mechanical Recycling

Both natural and synthetic fibers can be recycled mechanically, a process that involves chopping the fibers, blending them with virgin fiber, and respinning them to form new yarns. **Mechanical recycling** can be used for other post-industrial or post-consumer waste, and it begins with collection. The items are then separated by type and color, and chopped into small pieces (usually measuring 2–6 square inches). The fabric is then garneted, a process that chops the yarns and fabrics to a fibrous condition by running the material through a series of high-speed cylinders covered with wire (e.g. saw wire), or steel spikes, until it forms individual fibers. Since the waste is usually sorted by color prior to garneting, each resulting bale of recycled fiber is one color or one color family. The yarn spinner can create custom shades and heather effects by blending fibers from several different bales together, not unlike the way dyes are used to create tertiary colors. The visual effect can be quite striking, since colors created this way have a depth and liveliness.

Since the process of chopping shortens and weakens the fibers, it is necessary to blend the recycled fiber with virgin fiber to strengthen the yarn. Recycled cotton is often blended with virgin cotton or synthetic fiber to help facilitate processing and add strength to the yarn. A typical blend is around 85 percent recycled/15 percent virgin cotton or synthetic fiber. The synthetic fiber used is most often acrylic, because this adds softness to the yarn and the resulting fabric, although polyester (particularly recycled polyester) is becoming more common. By blending recycled cotton with recycled polyester, a 100 percent recycled product can be produced. It should be noted that a blend of cotton and polyester is neither biodegradable nor recyclable at this time.

Cotton, wool, polyester, and nylon fibers are most often mechanically recycled at this time.

Mechanically recycled fibers are considered "downcycled" because over time their quality naturally deteriorates. The process of chopping/grinding shortens and weakens the fiber, culminating into the need to eventually dispose of the fiber. At this time, chemical recycling fibers is suggested. For natural or synthetic fibers that cannot be chemically recycled, an interim solution is downcycling these fibers into industrial products, for example, rags, insulation for construction material, cushioning, and filling for stuffed toys. This type of downcycling offers a once-through solution, requiring incineration or energy recovery once discarded—so this is certainly not a permanent solution. The value is that these solutions displace virgin fiber use. Permanent solutions looking at the continuous cycling of textiles are recommended and currently being explored.

Mechanically recycled fiber can also reduce impacts associated with dyeing, since the color from the previous generation of the garment remains in the fiber, and avoids emissions associated with disposal in landfill. When they are overdyed, recycled fibers incur the usual toxicity, water, and energy impacts associated with dyeing processes.

- Visit **Chapter 23: Dyeing and Printing** for more guidance on these impacts.

## Melt Processing

Some manufactured fibers, particularly polyethylene terephthalate (PET), can be melt processed: the fiber can be effectively remelted and remolded to make yarns. However, in this manner the fiber is downcycled; its physical structure breaks down, and eventually the product must be discarded to landfill[1] or in Europe and Asia, sent to incineration with energy recovery.

Collection, sorting, and purifying discarded synthetic garments (i.e., post-consumer waste) is currently cumbersome. Infrastructure for labeling, collection, and sorting needs to be improved so that the post-consumer raw material source can scale to be economically viable.

Polymer resins come in a variety of forms, and some are relatively easy to collect and recycle. The most well-known source is soda bottles, which can be used to make new PET fiber. The bottles are collected; sorted by color (green versus clear); and thoroughly inspected to ensure that no caps (often polypropylene), bases, or PVC bottles are present. (This is critical, because one stray PVC bottle in a melt of 10,000 PET bottles can ruin the entire batch of new fiber.) Following inspection, the bottles are sterilized, dried, and crushed into flakes, which are washed again, bleached, and dried. The flakes are then emptied into a vat, heated, melted, and extruded through spinnerets to form long polyester fibers. Flakes from green bottles are generally used for fibers that will be dyed in dark colors, though some companies take advantage of the green color in the new fabric developed.

**Figure 26.1**
Cotton, wool, polyester, and nylon fibers are most often mechanically recycled at this time.
*Photo 12/UIG via Getty Images*

## Chemical Recycling

### Polyethylene Terephthalate

Chemical recycling involves breaking the polymer into its molecular parts and reforming the molecules into a yarn of equal strength and quality as the original, in perpetuity.[2] In this process, the chemical building blocks are separated (depolymerization) and reassembled (repolymerization), forming what is known as a "closed loop," where the final stage of the product's life cycle (disposal) forms the first stage of the next product (raw fiber). Closed-loop recycled polyester processing is expensive in part because it is a relatively new technology. In addition, the infrastructure to label, collect, sort, and purify discarded garments at scale is being developed.

In 2002, the Japanese company Teijin launched ECO CIRCLE™, the first closed loop chemical recycling system for polyester. Teijin works with fabric suppliers and apparel brands to manufacture products using recycled and recyclable materials, and it is helping to develop post-consumer clothing collection programs. Teijin recently established a joint venture with one of China's largest fiber producers, bringing the manufacture of chemically processed recycled polyester to China.[3]

There are many emerging technologies for the recycling of post-consumer textiles into new virgin quality textiles through chemical processing. Per the U.S. Environmental Protection Agency's 2012 report, 28 billion pounds of textiles is discarded by U.S. homes annually. That's about 70 pounds per person. Only 4.2 billion pounds (15 percent) of that is donated or recycled, leaving 85 percent or just under 24 billion pounds of textile waste going to U.S. landfills.

As a result of this discovery, many global companies have materialized to capitalize on this previously untapped resource of feedstock or input. These companies have developed their unique chemical recycling solutions to processing post-industrial and/or post-consumer waste into virgin-quality fibers. Two of these companies, Re:newcell and Evrnu, featured in the Future Fibers: Promoting Circular Textiles section, are both processing 100 percent cotton as input and regenerating into virgin-quality fibers. These fibers have the ability to be continuously recycled.

Another company in the United Kingdom, Worn Again, has partnered with Kering, Puma, and H&M to commercialize its technology of separating cotton/

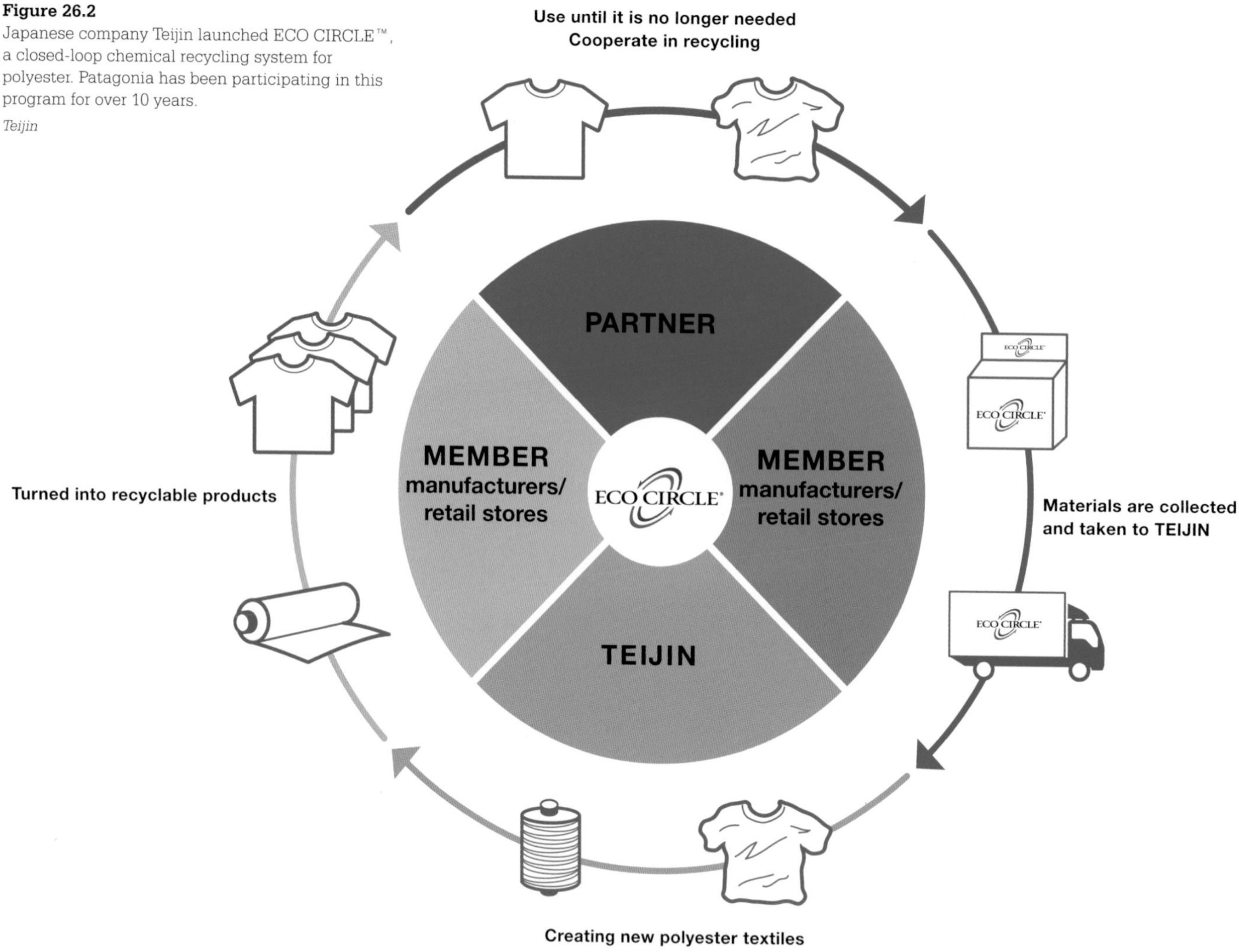

**Figure 26.2**
Japanese company Teijin launched ECO CIRCLE™, a closed-loop chemical recycling system for polyester. Patagonia has been participating in this program for over 10 years.
*Teijin*

Use until it is no longer needed
Cooperate in recycling

PARTNER

MEMBER
manufacturers/
retail stores

ECO CIRCLE®

MEMBER
manufacturers/
retail stores

TEIJIN

Materials are collected
and taken to TEIJIN

Creating new polyester textiles

Turned into recyclable products

polyester blends and creating both a regenerated fiber from the cotton and a recycled fiber from the polyester.

The Cradle to Cradle Products Innovation Institute, located in San Francisco, has gathered a collective of global innovative fiber recyclers/ manufacturers and apparel brands to facilitate the development of these new chemical fiber recycling technologies to allow for Cradle to Cradle solutions and innovations for pre-consumer and post-consumer waste/feedstock into new virgin-quality yarn.

- Visit **Chapter 23: Dyeing and Printing** for more information on impacts associated with dyeing.
- Visit **Appendix: Biodegradability in the Fashion Fibers STUDIO** for information about designing for biodegradability

# Optimize Sustainability Benefits

- Whatever the process, unsafe input becomes unsafe output. Contribute to a safe output by ensuring that the materials you use are safe for people and the environment

**Table 26.1: Comparison Chart Between Mechanical Recycling, Chemical Recycling, and Melt Processing**

| Process | Benefits | Considerations | Impacts |
|---|---|---|---|
| Mechanical recycling | • Slows the depletion of non-renewable resources.<br>• Fewer $CO_2$ emissions than virgin fiber.<br>• Diverts textile waste from landfills or waste-to-energy incineration (keeps fiber cycling and circumvents need to use virgin resources). | • Difficult to label, collect, sort, and purify post-consumer garments on a large scale.<br>• Some fabrics with chemical backing, lamination, or finish or those used in complex blends with other synthetics (nylon, for example) are not physically recyclable.[4]<br>• This process degrades the fiber and eventually the product is disposed of in a landfill.<br>• Beware: The demand for used PET bottles is now surpassing supply in some areas, and reports indicate that some suppliers are buying new bottles to make polyester textile fiber that can be called recycled.[5] | • In some cases re-dyeing will need to occur. Re-dyeing greatly increases levels of water, energy, and chemicals used.<br>• Whites can also be difficult to achieve in recycled fibers, and some processors use chlorine-based bleaches to whiten the base fabric. The dyeing and bleaching process for recycled fabrics involves standard industry chemicals. |
| Chemical recycling | • Slows depletion of nonrenewable resources.<br>• Generates fewer $CO_2$ emissions than virgin polyester.<br>• Diverts textile waste from landfills or waste-to-energy incineration (keeps fiber cycling and circumvents need to use virgin resources).<br>• Creates a completely new yarn of equal strength and quality to virgin fiber, in perpetuity. | • Post-consumer input is difficult to label, collect, sort, and purify on a large scale.<br>• Some fabrics with chemical backing, lamination, or finish or those used in complex blends with other synthetics are not chemically recyclable.[6] | • Uses significant amounts of energy.<br>• Chemicals used in the process and water discharge should be assessed and guaranteed "safe." |
| Melt processing recycling | • Slows depletion of nonrenewable resources.<br>• Generates fewer $CO_2$ emissions than virgin polyester.<br>• Diverts textile waste from landfills or waste-to-energy incineration (keeps fiber cycling and circumvents need to use virgin resources).<br>• Can create a completely new yarn of almost equal strength and quality to virgin fiber. | • Difficult to label, collect, sort, and purify discarded polyester garments on a large scale.<br>• Some fabrics with chemical backing, lamination, or finish or those used in complex blends with other synthetics are not chemically recyclable.[7] | • Melt processed fiber will have some degradation and therefore cannot be recycled more than a few times. |

- Use mechanically or chemically recycled polyester that is antimony free.
- Contribute to industry-wide collaboratives such as the Sustainable Apparel Coalition (SAC) and Cradle to Cradle's Fashion Positive Initiative that are developing systems for take-back and sorting. Participating in these initiatives will help the highlighted companies ensure they get feed-stock for these fibers.

**Figure 26.3**
Recycled polyester chip from mechanically recycled PET.
*Anne Green-Armytage/Getty Images*

# Potential Marketing Opportunities

- **X% recycled content:** If recycled content source has been verified.
- **Produced from textile industry waste:** If verified that the waste used for recycled fibers has been taken from what would otherwise be incinerated or landfilled.
- **Biodegradable:** All fibers, yarns, trims, and dyes used to manufacture the product or garment must also be biodegradable or disassembled before disposal. This should be substantiated with documentation that the product can completely break down into nontoxic material by being processed in a facility where compost is accepted. A secondary label or marketing material should be provided to instruct the customer. It is still to be determined whether Evrnu or Re:newcell will be proven to be biodegradable.

# INNOVATION EXERCISES

## Design

1. Design a garment that emphasizes one or more of the sustainability benefits of a recycled fiber.
2. Identify a potential environmental impact of one of the fibers detailed in this section. How would an intelligent design circumvent this impact?

## Merchandising

3. Create a fifty-word marketing communication message for consumers about the sustainability benefits of one of the following types of fibers: mechanical, chemical, or melt processed.
4. Incorporating information from chapters 1, 10, 11, and 23, develop a table that a retail buyer could use to quickly compare the advantages and

disadvantages of 100 percent cotton, 100 percent polyester, 100 percent nylon, and recycled fibers.

5. Create a retail point-of-purchase sign or hangtag to promote fashion merchandise made from recycled fibers.
6. Develop a marketing research protocol (e.g., focus group protocol, online survey, observational strategy) to determine consumers' perceptions of recycled fibers.

# REVIEW AND RECOMMENDATION

## Recommended Improvements

- Use fiber that has been *originally* dyed using environmental dye practices so the output fiber is clean and not contaminated.
- Incorporate environmental dyeing practices.
- Develop end-of-use strategy (whether compostability or recyclability).

James Hardy/PhotoAlto/Getty Images

# FUTURE FIBERS: PROMOTING CIRCULAR TEXTILES

Recently, a new category of recycled fibers, regenerated fibers, has sprung up globally to address the challenges of traditional mechanically, chemically, and melt-processed fibers. The definition of regenerate is to "revive or restore," and this section features companies that are currently developing regenerated virgin-quality fibers by restoring what we would consider to be waste. Regenerated fibers is a new category in the fashion industry, and thus they do not fall within traditional fiber categories. The fibers detailed in this chapter have been created through proprietary chemical processes and are addressing current challenges in the fashion industry, such as resource depletion/scarcity, $CO_2$ emissions, and landfill use.

These fibers are of particular interest to the fashion industry, as they are technically "recycled" fibers, but these processes can produce virgin-quality fibers, unlike traditional mechanical recycling methods for cotton.

# EVRNU [1]

Evrnu is a regeneration process that transforms post-consumer cotton waste into a virgin-quality regenerated fiber. The proprietary process separates dyes and contaminates and pulps the cotton, breaking it down to its constituent fiber molecules. These fibers are then recombined and extruded as a new fiber, which can be engineered to custom specifications, diameters, and cross-sectional shapes. The fiber characteristics most closely resemble silk in terms of hand and the ability to attract and hold color.

## Sustainable Benefits

Evrnu is turning millions of tons of garment waste into a renewable fiber resource. This fiber is thus contributing to reducing resource scarcity and $CO_2$ emissions from the creation of virgin fibers.

## Potential Impacts

Since Evrnu is currently in the research/pilot stage of development, it will need to be guaranteed that processing chemicals and water discharge are harmless and nontoxic.

Scaling up the technology post-research phase could potentially be a barrier.

## Biodegradability/Recyclability

Evrnu is in the infancy stages of determining the end use applications of garments made from Evrnu fiber; lab data indicates Evrnu garments can be broken down at least one additional time, but more testing is required to determine all suitable reuse options.

- Visit **Appendix: Biodegradability in the Fashion Fibers STUDIO** for information about designing for biodegradability.

## Availability

Evrnu is not yet available commercially, but will likely be available starting in 2017–2018.

## Fashion Applications

Evrnu can be engineered to create garments for a wide range of applications, including T-shirts, silk alternative applications, and denim.

**Figure P4.1**
The Evrnu process of regenerating cotton textile waste into virgin-quality fiber.
*Evrnu*

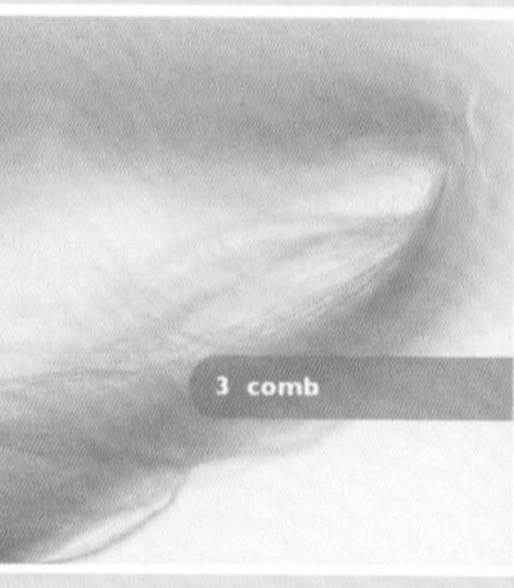

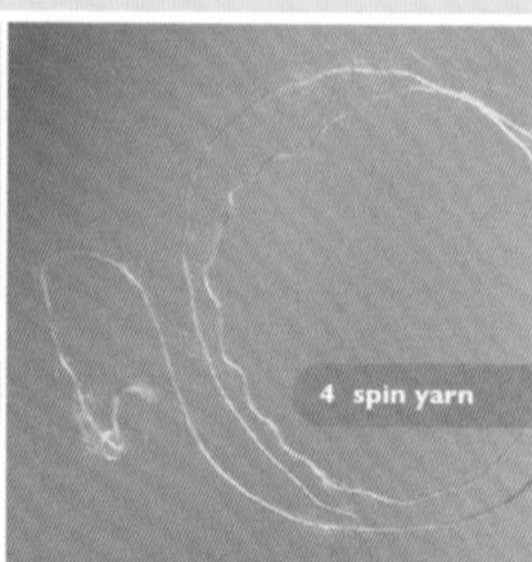

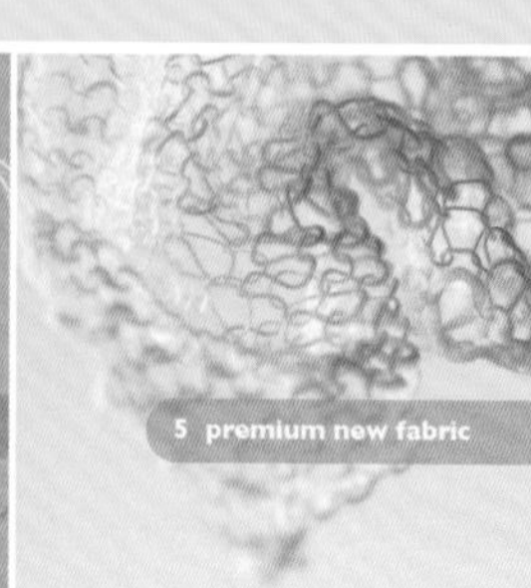

**Figure P4.2**
Cotton T-shirt produced from Re:newcell pulp.
*Re:newcell*

# Re:newcell[2]

Re:newcell process produces a pulp from post-industrial cotton waste, which can then be processed into regenerated yarn through the lyocell or viscose process.

### Sustainable Benefits

Like Evrnu, Re:newcell is capitalizing on fashion industry waste, in this case post-industrial. This fiber pulp is thus contributing to reducing resource scarcity and $CO_2$ emissions from the creation of virgin fibers.

Recent tests show that the fibers are of high quality in areas such as dyestuff absorption, tenacity in both wet and dry conditions, and withstanding high abrasion.

### Potential Impacts

Since Re:newcell is currently in the pilot stage of development, it will need to be guaranteed that processing chemicals and water discharge are harmless and nontoxic as well as potentially used in a closed-loop system of chemical use and recovery.

A barrier in scaling up is availability of raw material.

### Biodegradability/Recyclability

Biodegradability and recyclability are undetermined at this time.

**Part 4 Table 1 Other Circular Technologies to Product Regenerated Recycled Fibers**

| Company | Overview | Considerations | Availability | Applications |
|---|---|---|---|---|
| Worn Again (www.wornagain.info) | Worn Again has invented a textile-to-textile chemical recycling technology that is able to separate and extract polyester and cotton from pre-consumer and end-of-use clothing and textiles. Once separated, the recaptured polyester is restored and extruded into virgin-equivalent polyethylene terephthalate (PET) chips. The cellulose is extracted from cotton to make a new, circular cellulosic fiber. The technology and approach address major barriers in textile-to-textile recycling, namely how to separate blended-fiber garments and how to separate dyes and other contaminants from polyester and cellulose before reintroducing the raw materials back into the supply chain. Worn Again has partnered with Kering, PUMA, and H&M to bring its technology to commercialization. | Worn Again is currently in the pre-commercialization stage and working to refine its technology. | Currently in the research and development phase. | Same application types for virgin polyester and regenerated cellulose. |
| Ioniqa (www.ioniqa.com/) | Ioniqa has developed a circular process using Magnetic Smart materials that can recycle any type of PET materials into virgin-quality PET raw materials (monomers). Ioniqa's process is able to remove the color in a clean and economic way, resolving the biggest challenge in the PET recycling industry and closing the loop. | Ioniqa is currently in the pilot stage and working to refine its technology on a pilot scale. | Currently in the pilot phase. Fiber using Ioniqa's process is not yet commercially available. | Same applications as those for virgin polyester. |

### Availability

Re:newcell will be commercially available starting in late 2016.

### Fashion Applications

Re:newcell can be used in a variety of textile woven and knitted applications. Depending on the weight and construction of the cloth, these fabrics may be suitable for shirts, skirts, dresses, evening gowns, home furnishings, and bedding.

### Optimize Sustainability Benefits

- Encourage these companies to attain a certificate for chemical use, ensuring that only safe chemicals are used throughout the process. Certificates include Cradle to Cradle Certified and Bluesign approved.
- Whatever the process, unsafe input becomes unsafe output. Contribute to a safe output by ensuring that the materials you use are first-generation materials that are safe for people and the environment.
- Contribute to an industry-wide collaborative such as the Sustainable Apparel Coalition (SAC) and Cradle to Cradle's Fashion Positive Initiative, which are developing systems for take-back and sorting. Participating in these initiatives will help the highlighted companies ensure they get feedstock for these fibers.

### Potential Marketing Opportunities

- **Produced from textile industry waste:** If verified that the waste used for Re:newcell or Evrnu has been taken from what would otherwise be incinerated or landfilled.
- **Biodegradable:** All fibers, yarns, trims, and dyes used to manufacture the product or garment must also be biodegradable or disassembled before disposal. This should be substantiated with documentation that the product can completely break down into nontoxic material by being processed in a facility where compost is accepted. A secondary label or marketing material should be provided to instruct customer. It is to be determined whether Evrnu or Re:newcell will be proven to be biodegradable.
- **X% recycled content:** If recycled content source has been verified.

## INNOVATION EXERCISES

Design

1. Design a garment that emphasizes one or more of the sustainability benefits of a regenerated fiber.
2. Identify a potential environmental impact of one of the fibers detailed in this section. How would an intelligent design circumvent this impact?

Merchandising

3. Create a fifty-word marketing communication message for consumers about the sustainability benefits of Evrnu or Re:newcell.
4. Incorporating information from chapters 1, 10, and 11 and Future Fibers: Promoting Circular Textiles, develop a table that a retail buyer could use to quickly compare the advantages and disadvantages of 100 percent cotton, 100 percent polyester, 100 percent nylon, and regenerated fibers (Evrnu or Re:newcell).

5. Create a retail point-of-purchase sign or hangtag to promote fashion merchandise made from a regenerated fiber (Evrnu or Re:newcell).
6. Develop a marketing research protocol (e.g., focus group protocol, online survey, observational strategy) to determine consumers' perceptions of regenerated fibers (Evrnu or Re:newcell) for fashion merchandise.

## REVIEW AND RECOMMENDATION

### Recommended Improvements

- Incorporate environmental dyeing practices.
- Develop end-of-use strategy (whether compostability or recyclability).

# Appendix A

# Social and Cultural Sustainability

As human population and wealth per person increase, we tend to use more natural resources, transforming them into consumer goods for sale. The exponential growth in raw material use is also attributed to current business models, which demand companies grow exponentially each year by selling more units. Sustainability requires that humans learn to live and conduct business within the limits of the natural resources that are available for all species. There are many, many ways we can achieve this.

The bulk of this book investigates the choices designers have to mitigate the material and processing impacts of their products. It also presents the purchasing decisions consumers can make to support these designers in the marketplace. Making informed material choices is critical because what consumers and designers ask for and buy sends a message to the fashion industry, prompting brands, garment makers, weavers, spinners, and growers/extractors to adjust their work processes to market preferences.

But fashion isn't only the sum of materials and processing. It's also a lived social and cultural experience, lasting far beyond a purchase in a store. This cultural space of clothing is rarely explored or considered as a starting point for sustainability innovation because it falls outside the usual purview of brands whose main purpose is to sell more material units.

Yet each of us has at least one garment in our closet that we have kept for a long time. Why? Perhaps it reminds us of a particular person or a special time in our lives. We may have altered it to fit our own body type or have refrained from washing it because the fabric is delicate or ornamented. Garments are marked by these individual "use practices,"[1] which express our personal values and our unique abilities to adapt garments to our lives and ourselves and to connect with each other.

In a material sense, use practices make a valuable contribution to fashion and sustainability because they often "slow the flow" of virgin resources and energy through our closets (and by extension through the fashion system). A garment washed less reduces water and energy in that phase of its life cycle, for example.

Use practices also amplify our emotional and cultural experience of fashion.

As Kate Fletcher notes:

> Use practices rarely, if ever, require interventions by scientists, technicians or fashion experts to persist. Rather, they arise spontaneously and intuitively from everyday people as they experience their garments on a daily basis.[2]

An increasing number of designers are starting to consider "user-centered" aspects in their sustainable design process. Designing products to be altered easily by the wearer him/herself; constructing garments to be adaptable to different body types, thereby enabling sharing; choosing fabrics that acquire "tactile empathy"[3] with increased use; and developing items to evolve with the wearer over time through second and third life printing services[4] are just a few of these emergent concepts.

Keeping a garment in active, low-impact use for as long as possible (making alterations and repairs, sharing, leasing, reselling online, or donating—and buying—at thrift stores) and then physically recycling the material at the end of a very long and productive life ensures that the energy and natural resources embedded in the product are utilized over and over.

But more to the point, use practices see wearers as equal collaborators, working actively on solutions that both optimize the resources and energy embedded in our garments and intensify our emotional experience of clothing—and fashion as a whole.

Lynda Grose
Fashion and sustainability designer, author, educator, consultant

*Lynda has worked in the fashion industry for more than thirty years and has spent most of her career focused on sustainability issues. She co-founded ESPRIT* ecollection*, the first ecological line of clothing marketed internationally by a global corporation, and has worked with designers, artisans, farmers, and corporate executives to further sustainability in the fashion sector. She sees design as a tool to help give form to a sustainable society and is passionate about new roles for design in achieving this end.*

# Appendix B

# Consumer Care and Washing

The consumer use phase of a garment or product can significantly contribute to its overall environmental influence. While consumer care and washing may seem out of the company's control, using effective product design and marketing to educate consumers could significantly affect the influence that a garment or product could have on the environment.

## Environmental Impacts

*Laundry Detergent*

Certain at-home laundry detergents include ingredients that have been reported to have detrimental effects on humans and the environment.

- **Sodium lauryl sulfate (SLS)/sodium laureth sulfate (SLES):** Can cause irritation of the skin and eyes.[1]
- **1,4-dioxane:** Possible carcinogen, bioaccumulative in the environment, groundwater contaminate, nonbiodegradable.[2]
- **NPE (nonylphenol ethoxylate):** Persistent in the aquatic environment, moderately bioaccumulative, and extremely toxic to aquatic organisms.[3] Has been banned in Europe.[4]
- **Phosphates:** Difficult to remove from wastewater and often end up in rivers and lakes, where they cause algae blooms that negatively affect ecosystems and marine life. They have been banned in Europe for use in consumer detergents.[5]

Other ingredients including linear alkyl sodium sulfonates (LAS), petroleum distillates (a.k.a. naphthas), phenols, optical brighteners, sodium hypochlorite (bleach), EDTA (ethylene-diamino-tetra-acetate), and artificial fragrances have also been linked to various toxic effects on fish and animals, as well as allergic reactions in humans.[6]

*Dry-Cleaning*

Perchloroethylene (perc), the main chemical used in dry-cleaning, has been reported to have detrimental effects on humans and the environment and contributes to ozone depletion.

*Ironing*

Some fabrics, such as hemp and silk, wrinkle easily and require heavy pressing to render them smooth after washing. This can use significant amounts of energy over the long term.

## Suggestions for Consumers and Innovation Ideas

- Encourage the use of "phosphate free," "no bleach," "SLE free," and "NPE free" detergents.
- Encourage the use of biodegradable detergents since these tend to not contain the harmful ingredients.
- Encourage the use of plant- and animal-based ingredients, instead of petroleum based.
- Encourage the use of concentrated detergents. These have reduced packaging.
- Encourage washing and rinsing in cold water.
- Encourage spot cleaning.
- Suggest alternatives to dry cleaning with perc, including Solvon K4 and hydrocarbon solvents.
- Use fibers creatively and effectively to create garments or products that allow for less washing.
- Create a product where staining is intended as a design element, influencing the consumer to wash less.
- Create a garment that allows the consumer to detach and wash pieces of the garment that readily get soiled, saving on water.
- Design garments that utilize the natural wrinkling of the fabric as a design feature to influence customers to iron less and thus reduce use of energy.

Use the hangtag and labeling/point-of-sale (POS) marketing to communicate these suggestions and influence the consumer to take an active role in reducing environmental impacts.

# Glossary

*acid dye* A dye that bonds to a fiber via an ionic bond. Acid dyes are used on protein fibers such as silk and wool and may also be used on nylon.

*aliphatic solvents* Solvents that are composed of open-chain (noncyclic) hydrocarbon chains and are often highly flammable. Some are also neurotoxins. Examples include hexane, gasoline, and kerosene.

*antibacterial* Inhibiting the growth and development of bacteria.[1]

*antimicrobial* Inhibiting the spread of microorganisms.[2] Antimicrobial treatments for fashion provide solutions for odor caused by perspiration.

*antimony free* Antimony trioxide is a common catalyst for polyester and is also a suspected human carcinogen. Polyester that is catalyzed with an antimony-free alternative, such as titanium-based catalysts, is recommended.

*aromatic solvents* Solvents that are composed of ring-formation (cyclic) hydrocarbons and are often carcinogenic and highly flammable. Examples include benzene, toluene, and xylene.

*auxiliary chemicals* Chemicals used to assist in the dyeing process.

*basic dye* A cationic stain that will react with negatively charged material. It is not water soluble unless the base is converted to a salt. Basic dyes are strongly colored and bright but are not lightfast.

*Better Cotton Initiative* The Better Cotton Initiative (BCI) was established in 2005 and employs a holistic approach to cotton cultivation. It is an integrated program that reduces water and chemical use. BCI does not allow child, bonded/forced labor; incorporates better treatment of female workers and proper handling and training for the use of pesticides and fertilizers; and encourages a number of techniques that will reduce water use. BCI does not prohibit the use of genetically modified seed.[3]

*bioaccumulate* The accumulation of substances, such as pesticides, or other organic chemicals in an organism.[4]

*bio-based* Having been intentionally made from living or once-living organisms, typically after processing. Examples include rayon, polylactide (PLA), particle board, and casein.

*biodegradable* A fiber or garment that can be broken down into simpler substances by microorganisms, light, air, or water in a process that must be nontoxic.[5]

*biodiversity* An ecosystem containing a variety of species.

*biological* Derived from nature.

*biological oxygen demand (BOD)* Biochemical oxygen demand is a measure of the amount of oxygen used by microorganisms (e.g., aerobic bacteria) in the oxidation of organic matter. Natural sources of organic matter include plant decay and leaf fall. However, plant growth and decay may be unnaturally accelerated when nutrients and sunlight are overly abundant due to human influence. Urban and industrial runoff carries nutrients from lawn fertilizers and leaves, grass clippings, and paper from residential areas, which increases oxygen demand. Oxygen consumed in this process takes it away from other aquatic organisms that need it to live.[6]

*biologically efficient* A crop that requires few chemical inputs during the growing season, is mainly rain-fed, and can grow in diverse climates.

*cadmium* Heavy metal used as a PVC stabilizer for plastisol inks for screenprinting. If landfilled can contaminate groundwater.

*carbon neutral* A process or activity where net zero carbon emissions are achieved by balancing the amount of carbon released to the environment with the creation of renewable energy or "carbon offsetting" by paying others to remove the same amount of carbon gases from the environment that were created.

*cationic fixing agents* A class of dye fixing agents composed of positively charged ions. They strengthen ionic bonds that bond the dye to the fibers. Their health effects are still under investigation, but they can be toxic to animal, plant, and human life and may damage reproductive health.

*chemical oxygen demand (COD)* The standard method for indirect measurement of the amount of pollution (that can be oxidized biologically) in a sample of water.

*chemical recycling* Breaking the polymer into its molecular parts and reforming the molecule into a yarn of equal strength and quality as the original.

*chitosan* Chitosan is a linear polysaccharide material made by treating shrimp and other crustacean shells with sodium hydroxide. Ongoing research has found that when it is cured with UV light, the substance shows potential as an antimicrobial surface treatment.

*chlorine derivatives* Alternatives to chlorine, used mainly for fading pre-dyed and pre-printed items, including sodium or calcium hypochlorite or potassium permanganate.

*chromophoric* Any chemical group that produces color in a compound, such as the azo group.

*cellulose (fibers)* Natural and manufactured fibers generated from plants, such as cotton, linen, rayon/viscose, and modal.

*chronic* Lasting for a long period of time.

*climate change* The temperature of the earth's surface heating up at catastrophic rates due to human activity.

*closed loop process* Production system in which the waste or byproduct of one process or product is used to provide nutrients for nature or industry in making another product.[8]

*compostable* A product that is compostable is one that can be placed into a composition of decaying biodegradable materials and eventually turns into a nutrient-rich material.

*CRAiLAR* A fiber alternative to cotton that is derived from flax.

*cross-linking agent* A chemical substance used to link polymer chains together. Emulsion inks can be solvent based, although their use is rare in textile and garment printing.

*cultivation* The planting or harvesting of crops.

*culturally durable* The ability for a fashion garment to withstand changing seasonal trends.

*decompose* To become broken down into components; disintegrate.

*deforestation* Deforestation is the removing or clearing of a forest to include the cutting of all trees, mostly for agricultural or urban use. The remaining land is not reforested but is usually converted to a nonforest classification.[9]

*desertification* Desertification is the process that turns productive desert into nonproductive desert as a result of poor land management.[10]

*dew retted* Dew retting is the preferred way to extract bast fibers from the stalk of the plant. Dew retting utilizes the natural moisture of dew and takes over two to three weeks to break down the stems.

*direct printing* Printing techniques where the pigment or dye is applied directly to the textile.

*disperse dye* A dye that consists of microscopic non-water-soluble dye molecules dispersed in water that is then heated to 130°C in order to penetrate polyester fiber.

*dissolved oxygen (DO)* Oxygen is measured in its dissolved form as dissolved oxygen. If more oxygen is consumed than is produced, dissolved oxygen levels decline and some sensitive animals may move away, weaken, or die.[11]

*downcycling* Downcycling is the recycling of a material into a material of lesser quality. For example, the mechanical recycling of plastics, which turns them into lower grade plastics.

*durable water repellant (DWR)* A coating added to fabrics at the factory to make them water resistant (or hydrophobic).

*eco-aging* An alternative to sandblasting denim, eco-aging is a safer process to get a distressed look on denim. The process was developed by Fimatex Group in Italy and uses a vegetable mix composed of the waste from the food chain.

*emissions* Emissions is the term used to describe the gases and particles that are put into the air, or emitted, by various sources.[12]

*emulsion inks* Emulsion inks are used mainly for direct printing of fabrics and are typically based on aqueous dispersions (i.e., water based) of a binder and cross-linking agent. Emulsion inks can be solvent based, although their use is rare in textile and garment printing.

*enzyme retted* An alternative to traditional processes to extract bast fibers from the stalk of the plant using enzymes. Enzymes use less water than bleaching and are fully biodegradable.

*enzymes* Any of various proteins, as pepsin, originating from living cells and capable of producing certain chemical changes in organic substances by catalytic action (as in chemistry).

*Fair Trade* A global program that secures a minimum fiber price for the farmer aiming to cover the average costs of sustainable production.

*fast fashion* Fashion or products that are designed and produced in a short period of time and likely used and disposed of in a short period of time by the consumer.

*fast-growing natural resource* Plants for trees used for fiber that grow quickly and can be harvested at least once a year, including bamboo, flax, and hemp.

*fire resistant* Able to withstand exposure to a fire of a specified intensity for a specified period of time, usually one, two, or four hours.[13]

*fire retardant* Able to slow the spread of a fire.

*Forest Stewardship Council (FSC)* FSC is an independent, nonprofit organization that protects forests and sets standards under which forests and companies are certified.[14]

*genetically modified Bt cotton* The most successful variety of genetically modified cotton that has been engineered so that the genetic code of the plant includes a bacterial toxin (bacillus thuringiensis) that is poisonous to pests. This means that the crop comes under attack less often and therefore requires fewer pesticide sprays.

*Global Organic Textile Standard (GOTS)* Recognized as the world's leading processing standard for textiles made from organic fibers, GOTS defines high-level environmental criteria along the entire organic textiles supply chain and requires compliance with social criteria as well.[15]

*glutaraldehyde* A colorless, oily, liquid chemical with a pungent odor.[16]

*halogenated organic compounds (AOX)* AOX stands for "adsorbable organically bound halogens" expressed as chloride, and determined according to the relevant European Standard method. AOXs are substances that are adsorbed from water onto activated carbon. They may be volatile substances like trichloromethane (chloroform), chlorophenoles, and chlorobenzenes or complex organic molecules like dioxins and furans. Most AOXs are chlorine-containing molecules, but bromo- and iodo-AOXs may also occur.[17]

*herbicide tolerant* A form of genetic modification to create plants that are resistant to herbicides, allowing farmers to spray crops with herbicides without harming the crop plant.

*human carcinogen* A carcinogen is a chemical or physical agent that can cause cancer.[18] Carcinogens can be viruses, hormones, ionizing radiation, or solid materials.

*humectant* A substance that absorbs or helps another substance retain moisture, such as glycerol.

*hydrogen disulfide* Air emission caused by the process to produce rayon.

*hydrogen peroxide* A relatively safe chemical that decomposes into water and oxygen. About 90 percent of fabric bleaching carried out prior to dyeing/printing is performed using hydrogen peroxide.

*hydrogen sulfide* A colorless, flammable, hazardous gas used in leather tanning.

*indirect printing* Also called transfer printing, indirect printing is a process by which the colorant is applied first to a carrier material and then transferred to the textile via the application of heat and/or pressure.

*insoluble* Cannot be dissolved in a liquid.

*Integrated Pest Management (IPM)* Integrated Pest Management is an environmental approach that focuses on long-term prevention of pests by integrating biological control, habitat manipulation, and modification of cultural practices. Pesticides are used only after monitoring and established guidelines indicate pests exceed acceptable levels.[19]

*International Federation of Agriculture Movements (IFOAM)* IFOAM is the worldwide organization for the organic movement, uniting more than 800 affiliate organizations in over 100 countries. IFOAM maintains organic farming standards and organic accreditation and certification service.[20]

*irrigation* To supply (dry land) with water by means of ditches, pipes, or streams.

*lead* Heavy metal used as a PVC stabilizer for plastisol inks for screenprinting. If landfilled, it can contaminate groundwater.

*low water footprint in cultivation* Plants or trees that require very little water for cultivation, such as bamboo.

*lyocell* The generic name for a biodegradable fabric made out of treated wood pulp. TENCEL® is the brand name.

*manufactured cellulosic fibers* Fibers that have been derived from plants and chemically regenerated into types such as viscose/rayon and modal.

*manufactured protein fiber* Otherwise known as azlon, is composed of regenerated, naturally occurring protein derived from a number of sources, including soybean, peanut, casein (from milk), zein (from maize), and collagen/gelatin (from animal protein).

*manufactured synthetic fibers* Fibers that are created from chemicals and include types such as polyester, acrylic, and nylon.

*mechanical recycling (of polyester)* A recycling process in which used polyethylene terephthalate (PET) bottles and leftover materials from manufacturing processes are remelted and remolded to make yarns for recycled polyester. Mechanical recycling downcycles plastic into lower-grade plastics until they eventually end up in the landfill.[21]

*monomer* A molecule that may bind chemically or supramolecularly to other molecules to form a polymer.

*mordant* A mordant is a substance used to set dyes on fabrics or tissue sections by forming a coordination complex with the dye, which then attaches to the fabric or tissue.[22]

*motes* Small, immature seeds that may remain attached to cotton fibers.

*mutagens* A substance or preparation capable of inducing mutation.

*naphthol dye* True cold-water dyes, naphthol dyes may be used in ice water. They are used to dye cellulosic fibers and silk. Very popular for multicolored fabrics, as two colors may be placed side by side without bleeding. Very high toxicity; some colors are mutagens, carcinogens, teratogens, and tumorigens.

*nonbiodegradable* Cannot decompose naturally by living organisms or bacteria.

*nonrenewable resource* A resource that does not renew itself at a sufficient rate for sustainable economic extraction in meaningful human time frames.

*OEKO-TEX® certified* OEKO-TEX® is an independent, third party certifier that offers two certifications for textiles: OEKO-TEX® 100 (for products) and OEKO-TEX® 1000 (for production sites/factories). OEKO-TEX® 100 labels aim to ensure that products pose no risk to health. These products do not contain allergenic dyestuffs or dyestuffs that form carcinogenic aryl-amines and several other banned chemicals. The certification process includes thorough testing for a long list of chemicals.[23]

*organic* Disallows the use of genetically modified (GM) seeds and restricts or disallows the use of many synthetic agricultural chemicals.

*organic in-conversion cotton* Uses non-GM seed, is grown in the same manner as organic, but is still transitioning through this two-year phase to certification under the European organic standard. Once the land on which it is grown has completed the two-year requirement, the fiber can be labeled as organic.

*organochlorine* Synthetic organic compounds with chlorinated aromatic molecules.

*organotins* Compound for stabilizing PVC.

*oxygenated solvents* Solvents that are composed of molecules that contain oxygen. Examples include alcohols, glycol ethers, methyl and ethyl acetates, and ketones.

*ozone* Ozone gas is composed of three oxygen atoms ($O_3$) and is a much more powerful oxidizing agent than chlorine.

*plastisol inks* Plastisol inks are primarily used for direct and indirect (transfer) printing of garments and are typically vinyl resin (PVC) dispersed in plasticizer.

*polylactide (PLA)* Polylactide is mainly made from sugars derived from corn, though any abundantly available sugar, such as wheat, sugar beets, or sugarcane could also be used. PLA is a new class of polymer that is biodegradable under optimum conditions.

*polyvinyl chloride (PVC)* Synthetic plastic polymer.

*polyurethane laminate (PUL)* A polyurethane coating that is laminated onto fabrics such as polyester or cotton, and uses solvents in a chemical bonding process.

*post-consumer recycled content* Post-consumer recycled content is derived from used and discarded products (e.g., garments, carpet, and automotive upholstery) that are collected, deconstructed if necessary, and either recycled and used as raw material in the textile facility or sold and used for some other purpose unrelated to textiles and apparel.

*post-consumer waste* Includes used and discarded products (e.g., garments, carpet, and automotive upholstery) that are collected, deconstructed if necessary, and either recycled and used as raw material in the textile facility or sold and used for some other purpose unrelated to textiles and apparel. Post-consumer recycled cotton tends to be downcycled.

*post-industrial waste* Post-industrial waste (also known as pre-consumer waste) utilizes material created during product manufacturing. Examples of post-industrial waste include selvage from weaving, fabric remnants, cutting-room waste, and excess production, inventory, and unsold items.

*Programme for the Endorsement of Forest Certification (PEFC)* The Programme for the Endorsement of Forest Certification is an international nonprofit, nongovernmental organization dedicated to promoting sustainable forest management (SFM) through independent third-party certification.[24]

*propylene* ($C_3H_6$) An unsaturated organic compound that has one double bond and is the second simplest member of the alkene class of hydrocarbons.

*protein fiber* Protein fibers are derived from animals and include alpaca, leather, silk, and wool.

*porous* Porous fibers, such as bamboo, have small spaces in their structure that allow liquid and air to pass. This feature makes these fibers naturally breathable and moisture wicking.

*PU split (PU coated split leather)* PU coated split leather, also known as "PU split," refers to the lower layer of the hide that is split from the top layer, which is of the highest quality, and is considered pure leather. The lower layer, called "split," is applied in a thin layer of polyurethane (PU) with foil or extrusion that hardens on top. PU coated split leather is not considered 100 percent leather.

*rapidly renewable* A natural resource that can grow quickly and can be harvested at least once a year.

*reactive dye* A dye that attaches to the fiber chemically. Reactive dyes are primarily used on natural fibers (cotton, silk, wool, etc.) and rayon.

*retting* A process that separates bast fiber (e.g., jute, flax, hemp, and bamboo) from the stems using microorganisms and moisture. This is carried out in the field (as with dew retting) or in tanks (water or chemical retting).

*polyethylene* ( $(C_2H_4)_n$ ) The most common type of plastic; its primary use is in packaging.

*Roundup Ready®/Herbacide tolerant (Ht) cotton* Cotton seed, developed by Monsanto, that contains a gene not normally found in cotton plants. The gene allows the plants to withstand Roundup® herbicide. That means that cotton growers can spray their fields with Roundup® herbicide to control weeds without damaging the cotton plants.[25]

*recyclable* To pass through a cycle again; repeat a process from the beginning.

*renewable natural resource* Natural resources that can be replaced or replenished by natural processes or human action.

*sanforizing* Sanforizing is a controlled compressive shrinkage process, which is applied on woven fabric to achieve shrinkage before making the garments. After sanforizing, the residual shrinkage of woven fabric may be zero.[26]

*scouring* Scouring removes top finishes, largely accomplished by rinsing with an appropriate detergent.

*sizing agent* Sizing agents are applied to warp yarns to improve the strength of the yarn for weaving.

*Sustainable Cotton Project's (SCP) Cleaner Cotton* Sustainable Cotton Project's Cleaner Cotton Program takes a collaborative, holistic approach to growing cotton. The cotton grown in this program is called "Cleaner Cotton."

*sericin* A viscous gelatinous protein that forms on the surface of raw-silk fibers.

*staple* Average length of a spinnable fiber.

*stain repellent* Something that, when exposed to liquid (that could permanently stain a garment) has a reaction that causes the liquid to move away from instead of get soaked into the fiber.

*strike off* A large sample of printed fabric used to test production methods. A strike off is usually multiple yards and tests pattern registration, repeat, and color matching to the original design.

*sulfur dye* A water-insoluble dye used for cotton with high toxicity. Available in brown, black, and dark blue.

*surfactant* Substance that lowers the surface tension between two liquids or a liquid and a solid (e.g., emulsifiers, detergents, dispersants).

*tannin* An acidic chemical compound that prevents decomposition and often imparts color and is used in the leather tanning process to convert putrescible skin into nonputrescible leather.

*tanning* The process by which animal skins are rendered soft, pliable, and durable enough for clothing and accessories.

*teratogens* A chemical capable of interfering with the development of a fetus, causing birth defects.[27]

*transitional cotton* Refers to cotton being grown according to organic standards, but that has not yet been following those standards long enough to be certified.

*United States Environmental Protection Agency (EPA)* An agency of the U.S. federal government that was created for the purpose of protecting human health and the environment by writing and enforcing regulations based on laws passed by Congress.

*United States Federal Trade Commission (FTC)* An independent agency of the U.S. government that promotes consumer protection.

*upcycling* To process (used goods or waste material) so as to produce something that is of higher quality than the original.[28]

*vat dye* A dye that is insoluble in water and is reduced with a basic solution in order to activate the dye. Vat dyes are primarily used on cellulosic fibers due to the extremely high pH, but not exclusively.

*volatile organic compounds (VOCs)* Volatile organic compounds (VOCs) are emitted as gases from certain solids or liquids. VOCs include a variety of chemicals, some of which may have short- and long-term adverse health effects.[29]

*volatility* In the context of chemistry, refers to the tendency of a substance to vaporize.

*water repellent* Something that, when exposed to water or liquid, has a reaction that causes the liquid to move away from instead of get soaked into the fiber.

# Bibliography


## Chapter 1

*Agriculture Marketing Resource Center, Cotton Profile*. March 2012. http://www.agmrc.org/commodities_products/fiber/cotton-profile/ (accessed March 16, 2015).

Alter, Alexandra. "Yet Another 'Footprint' to Worry About: Water." *Wall Street Journal*. February 17, 2009. http://www.wsj.com/news/articles/SB123483638138996305 (accessed March 16, 2015).

Bayer, Alicia. *Examiner.com, What countries have banned GMO crops?* June 18, 2011. http://www.examiner.com/article/what-countries-have-banned-gmo-crops (accessed March 19, 2015).

*Better Cotton Initiative, BCI History*. http://bettercotton.org/about-bci/bci-history/ (accessed March 22, 2015).

Bishek, D. T. *The Economist, Forced labour in Uzbekistan: In the Land of Cotton*. October 16, 2013. Forced labour in Uzbekistan In the land of cotton (accessed March 17, 2015).

Bloch, Michael. *Green Living Tips, How long does X take to break down?* May 20, 2013. http://www.greenlivingtips.com/articles/waste-decomposition-rates.html (accessed March 17, 2015).

*Bureau Veritas, Organic Cotton Frequently Asked Questions*. 2010. http://www.bureauveritas.com/wps/wcm/connect/a42d00004261a59dbce0bf5744846dd0/Bureau+Veritas OrganicCotton_FAQ.pdf?MOD=AJPERES&lmod=- (accessed March 22, 2015).

*CUESA. From Farm to Garment: Growing a Sustainable Fibershed*. October 26, 2012. http://www.cuesa.org/article/farm-garment-growing-sustainable-fibershed (accessed March 16, 2015).

East, Roger. *Triple Pundit, What Does the Future Hold for Genetically Modified Cotton?* August 2, 2013. http://www.triplepundit.com/2013/08/future-hold-genetically-modified-cotton/ (accessed March 19, 2015).

*European Commission on Environment, Biodegradable Waste*. March 3, 2015. http://ec.europa.eu/environment/waste/compost/ (accessed March 17, 2015).

Glausiusz, Joise. *One Earth, How Green Are Your Jeans?* December 1, 2008. http://archive.onearth.org/article/how-green-are-your-jeans (accessed March 16, 2015).

Huang, Jikun, Ruifa Hu, Fan Cuihui, E. Pray Carl, and Scott Rozelle. "Bt Cotton Benefits, Costs, and Impacts in China." *Journal of Agrobiotechnology Management & Economics*, 2002: 153–166.

Mirovalev, Mansur, and Andrew E. Kramer. *New York Times, In Uzbekistan, the Practice of Forced Labor Lives On During the Cotton Harvest*. December 13, 2013. http://www.nytimes.com/2013/12/18/world/asia/forced-labor-lives-on-in-uzbekistans-cotton-fields.html?_r=0 (accessed March 17, 2015).

"Natural Fibers Cotton." *Discover Natural Fibers*. http://www.naturalfibers2009.org/en/fibers/cotton.html (accessed March 22, 2015).

"OE Blended Standard." *Textile Exchange*. 2009. http://textileexchange.org/sites/default/files/te_pdfs/integrity/OEBlended%20Standard_v1.3.pdf (accessed March 23, 2015).

"Oeko-Tex S." *Oeko-Tex*. https://www.oeko-tex.com/media/downloads/Factsheet_OETS_100_EN.pdf.

"Organic Trade Association." *Organic Cotton Facts*. June 29, 2012. https://ota.com/sites/default/files/indexed_files/Organic-Cotton-Facts.pdf (accessed March 22, 2015).

*Pesticide Action Network, Cotton*. http://www.panna.org/resources/cotton (accessed March 16, 2015).

Stochaj, Mariusz. "The Carbon Footprint of a Cotton T-Shirt." *Contintntal Clothing Co. Ltd*. March 24, 2009. files.continentalclothing.com/press/LCA 20Executive 20Summary.doc (accessed March 22, 2015).

*Sustainable Cotton Project, Better Cotton Field Program*. http://www.sustainablecotton.org/pages/show/cleaner-cotton-field-program (accessed March 22, 2015).

*Sustainable Cotton Project, Cleaner Cotton*. December 4, 2014. http://www.sustainablecotton.org/pages/show/cleaner-cotton-tm (accessed March 22, 2015).

"Textile Exchange, The global sustainable cotton situation." *Textile Exchange*. http://www.textile-future.com/dynpg/print_text.php?lang=en&aid=1476&showheader=N (accessed March 22, 2015).

*US Department of Agriculture Economic Research Service Cotton and Wool Overview*. October 17, 2014. http://www.ers.usda.gov/topics/crops/cotton-wool.aspx (accessed March 16, 2015).

*US Department of Agriculture, Adoption of Genetically Engineered Crops in the U.S*. August 26, 2014. http://www.ers.usda.gov/data-products/adoption-of-genetically-engineered-crops-in-the-us.aspx (accessed March 19, 2015).

Wakelyn, Phil. *Cotton's Revolutions*. http://cottonsrevolutions.org/applications/blog/blogentryview.cfm?itemnumber=3299 (accessed March 23, 2015).

*What is Oeko-Tex?*

*World Wildlife Fund, Cotton Farming*. 2015. http://wwf.panda.org/about_our_earth/about_freshwater/freshwater_problems/thirsty_crops/cotton/ (accessed March 16, 2015).

## Chapter 2

Bloch, Michael. 2013. *Green Living Tips, How long does X take to break down?* May 20. http://www.greenlivingtips.com/articles/waste-decomposition-rates.html (accessed March 17, 2015).

Food and Agriculture Organization of the United Nations. n.d. "Future Fibers." *Food and Agriculture Organization of the United Nations*. http://www.fao.org/economic/futurefibers/fibers/jute/en/ (accessed March 27, 2015).

Masters of Linen. n.d. "Be Linen Map." *Masters of Linen*. http://www.knittingindustry.com/uploads/2048/BE_LINEN_MAP_10-1.pdf (accessed March 27, 2015).

———. 2008. "The Linen Shirt Eco-Profile." *Masters of Linen*. February. http://www.saneco.com/IMG/pdf/linen_shirt_eco-profile.pdf (accessed March 27, 2015).

Parideh, Md. Tahir, Amel B. Basher, Syeed Saiful Azry, and Zakiah Ahmed. 2011. "Retting Process of Some Bast Plant Fibers and its Effects on Fiber Quality: A Review." *BioResources* 5260–5281. (accessed March 27, 2015).

## Chapter 4

Britannica, Encylopaedia. n.d. *Retting Fibre-separation process*. http://www.britannica.com/EBchecked/topic/500159/retting (accessed March 31, 2015).

Food and Agriculture Organization of the United Nations; Future Fibres. 2009. *Hemp*. http://www. naturalfibres2009.org/en/fibres/hemp.html (accessed March 31, 2015).

*Hemp Industry Association*. 2014. http://www.thehia.org/facts.html (accessed March 31, 2015).

Small, E., and D. Marcus. 2002. "Hemp: A New Crop with New Uses for North America." Edited by J. Janick and A. Whipkey. *Proceedings of the Fifth National Symposium: Trends in New Crops and New Uses*. Alexandria: ASHS Press. 284–326. https://www.hort.purdue.edu/newcrop/ncnu02/v5-284.html (accessed March 31, 2015).

## Chapter 5

AVM Chemical. n.d. "What Is Jute? Some Features of Jute: Jute Processing & Their Uses." *AVM Chemical*. http://avmchemical.com/article/ (accessed June 3, 2015).

Bloch, Michael. 2013. May 20. http://www.greenlivingtips.com/articles/waste-decomposition-rates.html.

Food and Agriculture Organization of the United Nations. n.d. *Future Fibres*. http://www.fao.org/economic/futurefibres/fibres/jute/en/ (accessed March 26, 2015).

Inagaki, H (2000). Progress on Kenaf in Japan. Third Annual Conference, held at American Kenaf Society, Texas, USA, 2000.

Lam Thi Bach Tuyet, Hori Keko and Iiyama Kenzi (2003). *Journal of Wood Science* 49(3): 255-261.

Li, Lili, Margaret Frey, and Kristie J. Browning. 2010. "Biodegradability Study on Cotton and Polyester." *Journal of Engineered Fibers and Fabrics* 42–53.

Mussig (Editor), Jorg. 2010. *Industrial Applications of Natural Fibres: Structures, Properties, and Tehnical Applications*. John Wiley & Sons, Ltd.

New York State Department of Health. 2004. *The Facts About Ammonia*. July 28. http://www.health.ny.gov/environmental/emergency/chemical_terrorism/ammonia_general.htm (accessed March 26, 2015).

PPG Industries, Inc. n.d. "NaOH." *Caustic*. http://www.ppg.com/chemicals/chloralkali/products/Documents/CausticSodamanual2008.pdf.

———. 2008. "NaOH Caustic Soda." Accessed March 26, 2015. http://www.ppg.com/chemicals/chloralkali/products/Documents/CausticSodamanual2008.pdf.

Pure Jute. n.d. *Jute and environment*. Accessed March 26, 2015. http://www.purejute.com/en/pure-jute/jute-environment.html.

Tahir, Paridah Md., Amel B. Ahmed, Syeed S. O. Saiful Azry, and Zakiah Ahmed. 2011. "https://www.ncsu.edu/bioresources/BioRes_06/BioRes_06_4_5260_Paridah_ASZ_Retting_Bast_Fiber_Quality_Review_1312.pdf." *Bioresources.com* https://www.ncsu.edu/bioresources/BioRes_06/BioRes_06_4_5260_Paridah_ASZ_Retting_Bast_Fiber_Quality_Review_1312.pdf (accessed March 26, 2015).

*YouTube, Jute Production*. July 17, 2012. http://www.youtube.com/watch?v=pIhRw1V0b60 (accessed March 27, 2015).

## Chapter 6

Benedictus, Leo. 2013. "Can angora production ever be ethical?" *The Guardian*, December 16. http://www.theguardian.com/world/shortcuts/2013/dec/16/angora-production-ethical-peta-video-chinese-rabbits (accessed March 29, 2015).

"Biological Oxygen Demand (BOD) – Overview." n.d.http://www.polyseed.com/misc/BODforwebsite.pdf (accessed March 29, 2015).

Bloch, Michael. 2013. *Green Living Tips, How long does X take to break down?* May 20. http://www.greenlivingtips.com/articles/waste-decomposition-rates.html (accessed March 17, 2015).

"Chemicals in the Environment: Chlorine." 1994. *U.S. Environmental Protection Agency*. August. http://www.epa.gov/chemfact/f_chlori.txt (accessed March 29, 2015).

"China sees growing demand for 'soft gold' cashmere." 2011. *Dawn.com*. May 4. http://www.dawn.com/news/625980/china-sees-growing-demand-for-soft-gold-cashmere (accessed March 30, 2015).

Fisher, Andrew D. 2011. "Addressing pain caused by mulesing in sheep." *Applied Animal Behaviour Science* 232-240.

*Food and Agriculture Organization of the United Nations*. 2009. http://www.naturalfibers2009.org/ (accessed March 30, 2015).

Gillet, Kit. 2011. *CNN, Cost of Cashmere in Mongolia*. September 13. http://www.cnn.com/2010/WORLD/asiapcf/09/12/mongolia.cashmere.herders (accessed March 29, 2015).

King, Bart. 2013. "Patagonia Cultivating Sustainable Wool in Namesake Region." *Sustainable Brands*. January 31. http://www.sustainablebrands.com/news_and_views/articles/patagonia-cultivating-sustainable-wool-namesake-region (accessed March 30, 2015).

*Novartis Animal Health Australia*. 2015. http://ah.novartis.com.au/livestock_products/clik.html/section/470 (accessed June 3, 2015).

O Eco Textiles. 2009. *What does organic wool mean?* http://oecotextiles.wordpress.com/2009/08/11/what-does-organic-wool-mean/ (accessed March 29, 2015).

Russell, I. M. n.d. "Sustainable Wool Production and Processing." In *Sustainable Textiles Lifecycle and Environmental Impact*, by R. S. Blackburn, 63–87. Cambridge: Woodhead Publishing.

*What Makes Wool So Special?* n.d. http://www.woolrevolution.com/virtues.html (accessed June 3, 2015).

Zhang, Wei, Tie-Ling Xing, Qiang-Hua Zhang, and Guo-Qiang Chen. 2014. "Thermal Properties of Wool Fabric Treated by Phosphorous-Doped Silica Sols." *Thermal Science* 1603–1605.

## Chapter 7

Bloch, Michael. 2013. *Green Living Tips, How long does X take to break down?* May 20. http://www.greenlivingtips

.com/articles/waste-decomposition-rates.html (accessed March 17, 2015).

European Commission. n.d. *Agriculture and Rural Development*. http://ec.europa.eu/agriculture/quality/schemes/index_en.htm (accessed March 28, 2015).

Evripidis, Kipriotis. n.d. *The prospects of the European Sericulture within the frame of the EU Common Agricultural Policy*. http://www.bacsa-silk.org/en/the-prospects-of-the-european-sericulture-within-the-frame-of-the-eu-common-agricultural-policy/ (accessed March 28, 2015).

Kulkarni, Vishwanath. 2013. "Govt may use wastelands for tasar silk cultivation." *The Hindu Business Line*. March 5. http://www.thehindubusinessline.com/industry-and-economy/agri-biz/govt-may-use-wastelands-for-tasar-silk-cultivation/article4479274.ece (accessed March 28, 2015).

LenTech. n.d. *Health Effects of Tin*. http://www.lenntech.com/periodic/elements/sn.htm#Health%20effects%20of%20tin (accessed March 28, 2015).

McLaren, Warren. 2006. *TreeHugger, Q&A: Is Silk Green?* October 20. http://www.treehugger.com/culture/qa-is-silk-green.html (accessed March 28, 2015).

Slater, K. 2003. *Environmental Impact of Textiles: Production, Processes and Protection*. Cambridge: Woodhead Publishing.

*Textile Exchange*. n.d. http://textileexchange.org/node/1096 (accessed March 28, 2015).

United Nations Food and Agriculture Organization. 2009. *Silk*. http://www.naturalfibres2009.org/en/fibres/silk.html (accessed March 28, 2015).

*WildSilkbase*. n.d. http://www.cdfd.org.in/wildsilkbase/info_moths.php (accessed March 28, 2015).

**Chapter 8**

Dima Nazer, Rashad Al-Sa'ed, and Maarten Siebel. 2006. "Reducing the environmental impact of the unhairing-liming process in the leather industry." *Journal of Cleaner Production* (Elsevier) 14: 65–74. http://www.academia.edu/1434007/Reducing_the_environmental_and_economic_impact_of_the_unhairing-liming_process_in_the_leather_tanning_industry (accessed March 11, 2015).

Mwinyihija, M. 2010. *Ecotoxicological Diagnosis in the Tanning Industry*, Springer Science+Business Media, LLC. doi:10.1007/978-1-4419-6266-9_2 (accessed March 11, 2015).

**Chapter 9**

Bloch, Michael. 2013. May 20. http://www.greenlivingtips.com/articles/waste-decomposition-rates.html (accessed March 11, 2015).

*What is Oeko-Tex?* n.d. https://www.oeko-tex.com/en/consumers/what_is_oeko_tex/what_is_oeko_tex.html (accessed March 11, 2015).

Ziek, Barbara. 2012. *What's So Special About Alpaca Fiber?* April 11. http://www.wildhairalpacas.com/pages/1414/wild-hair-alpacas-llc-whats-so-special-about-alpaca-fiber (accessed March 11, 2015).

**Chapter 10**

"AirDye Environmental Profile Life Cycle Assessment." n.d. *AirDye*. http://www.airdyesolutions.com//uploads/AirDye_EPDv2b_091109.pdf (accessed April 23, 2015).

Boh, Richard, interview by Annie McCourt Gullingsrud. 2014. (February 25).

Braun, Emil, and Barbara C. Levin. 1986. "Polyester: A Review of the Literature on Products of Combustion and Toxicity." *Fire and Materials* (US National Bureau of Standards) 10: 107–123. http://fire.nist.gov/bfrlpubs/fire86/PDF/f86012.pdf (accessed April 25, 2015).

Coen, Amanda. 2011. *Ecouterre*. October 24. http://www.ecouterre.com/is-synthetic-clothing-causing-microplastic-pollution-in-oceans-worldwide/ (accessed April 21, 2015).

Ducas, Charlene. 2012. "Athleta Webinar: Textile Fibers & Sustainability." October 29.

*European Commission, EU Crude Oil Imports*. 2001–2014. http://ec.europa.eu/energy/observatory/oil/import_export_en.htm (accessed April 25, 2015).

Grose, Linda, and Kate Fletcher. 2012. *Fashion and Sustainability: Design for Change*. London: Laurence King Publishing, Ltd.

King, David. 2010. "Top scientist says politicians have 'heads in the sand' over oil." *The Guardian*, June 9. http://www.guardian.co.uk/environment/2010/jun/09/sir-david-king-dwindling-oil-supplies (accessed April 21, 2015).

"Locating and Estimating Air Emissions from Sources of Acrylonitrile." 1984. *U.S. Environmental Protection Agency*. March. http://www.epa.gov/ttnchie1/le/acrylon.pdf (accessed April 19, 2015).

*Make Wealth History, How much oil is there left, really?* n.d. http://makewealthhistory.org/2010/06/11/how-much-oil-is-there-left-really/ (accessed April 11, 2015).

Mowbray, John. 2008. *EcoTextile News*. May 13. http://www.ecotextile.com/20080513796/materials-production-news/concern-over-recycled-polyester.html (accessed April 21, 2015).

*New York Fashion Center, Polyester Fabric Information*. n.d. http://www.nyfashioncenterfabrics.com/polyester-fabric-info.html (accessed April 21, 2015).

*O Ecotextiles*. n.d. http://oecotextiles.wordpress.com/2009/07/14/why-is-recycled-polyester-considered-a-sustainable-textile/#_ftn6 (Accessed April 23, 2015).

"Oeko-Tex Standard 100." n.d. *Oeko-Tex*. https://www.oeko-tex.com/media/downloads/Factsheet_OETS_100_EN.pdf (accessed April 5, 2015).

Ryan, V. n.d. *Technology Student*. http://www.technologystudent.com/joints/pla1.html (accessed April 25, 2015).

Scaturro, Sarah. 2011. *Ecouterre*. March 1. http://www.ecouterre.com/could-polyester-be-the-next-eco-friendly-fabric/ (accessed April 21, 2015).

Sundar, Shyam, and Jaya Chakravarty. 2010. "Antimony Toxicity." *Int J Environ Res Public Health*. 4267–4277. doi:10.3390/ijerph7124267 (accessed April 21, 2015).

*Swicofil, Polyester*. 2013. http://www.swicofil.com/pes.html (accessed April 21, 2015).

*Teijin, Closed Loop Recycling System ECO CIRCLE*. n.d. http://www.teijin.com/solutions/ecocircle/ (accessed April 23, 2015).

"The Fiber Year 2009–2010." 2010. *Oerlikon Textile*. May. http://www.indotextiles.com/download/Fiber%20Year%202009_10.pdf (accessed April 21, 2015).

*Victor*. n.d. http://www.victor-innovatex.com/en/ (accessed April 25, 2015).

## Chapter 11

*Aquafil Group Worldwide*. n.d. http://www.aquafil.com/en/about-us/worldwide.html (accessed April 25, 2015).
"Chemicals Released During Open Burning." 2005. *South Dakota Department of Environment and Natural Resources*. December 12. http://denr.sd.gov/des/wm/sw/documents/OpenBurningChemicalList.pdf (accessed April 19, 2015).
"Concern over Recycled Polyester." 2008.*The Textile Dyer*. May 13.
*Ensinger, Polyamides*. n.d. http://www.ensinger-online.com/en/materials/engineering-plastics/polyamides/ (accessed April 19, 2015).
*Eur Lex*. May 15, 2014. http://eur-lex.europa.eu/legal-content/EN/TXT/?uri=CELEX:32014D0520%2801%29 (accessed April 19, 2015).
Freinkel, Susan. 2011. *Plastic: A Toxic Love Story*. New York: Houghton Mifflin Harcourt.
Grose, Linda, and Kate Fletcher. 2012. *Fashion and Sustainability: Design for Change*. London: Laurence King Publishing, Ltd.
Hegde, Raghavendra , Atul Dahiya, and M.G. Kamath. 2004. "Nylon Fibers." *Nonwovens Science and Technology II*. April. http://www.engr.utk.edu/~mse/Textiles/Nylon%20fibers.htm.
"Locating and Estimating Air Emmissions from Sources of Acrylonitrile." 1984. *U.S. Environmental Protection Agency*. March. http://www.epa.gov/ttnchie1/le/acrylon.pdf (accessed April 19, 2015).
Mitchell, Carey. n.d. "Are There Real Differences Between Type 6 and 6.6 Nylons?" http://www.baneclene.com/articles/type6vstype66nylon.pdf (accessed April 25, 2015).
*O Ecotextiles*. n.d. http://oecotextiles.wordpress.com/2009/07/14/why-is-recycled-polyester-considered-a-sustainable-textile/#_ftn6 (accessed April 23, 2015).
"Oeko-Tex Standard 100." n.d. *Oeko-Tex*. https://www.oeko-tex.com/media/downloads/Factsheet_OETS_100_EN.pdf (accessed April 5, 2015).
*Patagoina*. March 4, 2009. http://www.thecleanestline.com/2009/03/closing-the-loop-a-report-on-patagonias-common-threads-garment-recycling-program.html (accessed April 25, 2015).
"Technical Bulletin." 2013. *Antron*. http://antron.net/na/pdfs/literature/K02510_N66vsN6_Tech_Bulletin_06_18_13.pdf (accessed April 19, 2015).
Upham, B.C. 2009. *AirDye: Dyeing Fabric Without Water*. July 24. http://www.triplepundit.com/2009/07/airdye-dyeing-fabric-without-water/ (accessed April 5, 2015).
Yamasaki, Yoshikazu. n.d. "Overview of Recycling Technology in Textile Industry in Japan and the World." http://hrd.apec.org/images/a/aa/62.4.pdf (accessed April 25, 2015).

## Chapter 14

*Composition Leather*. n.d. http://www.compositionleather.com/glossary (accessed April 12, 2015).
*European Commission Environment*. March 25, 2015. http://ec.europa.eu/environment/chemicals/endocrine/definitions/endodis_en.htm (accessed April 12, 2015).
*European Commission Scientific Committees Glossary*. n.d. http://ec.europa.eu/health/scientific_committees/opinions_layman/perfume-allergies/en/glossary/pqrs/phthalates.htm (accessed April 14, 2015).
*Healthy Child Healthy World*. January 30, 2013. http://healthychild.org/easy-steps/avoid-phthalates-find-phthalate-free-products-instead%E2%80%A8%E2%80%A8/ (Accessed April 14, 2015).
"Phthalates." *US Environmental Protection Agency*. October 10, 2007. http://www.epa.gov/teach/chem_summ/phthalates_summary.pdf (accessed April 14, 2015).
"Phthalates entry 52." *European Commission*. January 15, 2014. http://ec.europa.eu/enterprise/sectors/chemicals/files/reach/entry-52_en.pdf (accessed April 14, 2015).
*Toray Innovation by Chemistry*. 2013. http://www.ultrasuede.com/about/responsibly_engineered.html (accessed April 14, 2015).
*US Consumer Product Safety Commission, Phthalates*. n.d. http://www.cpsc.gov/en/Business—Manufacturing/Business-Education/Business-Guidance/Phthalates-Information/ (accessed April 14, 2015).

## Chapter 16

Browne, Mark A., Richard Thompson, Phillip Crump, Stewart J. Niven, Emma Teuten, Andrew Tonkin, and Tamara Galloway. 2011. "Accumulation of Microplastic on Shorelines Woldwide: Sources and Sinks." *Environmental Science and Technology* 45 (21): 9175–9179. doi:10.1021/es201811s (accessed April 25, 2015).
Coen, Amanda. 2011. *Ecouterre*. October 24. http://www.ecouterre.com/is-synthetic-clothing-causing-microplastic-pollution-in-oceans-worldwide/ (accessed April 212011, 2015).
"Composites Design and Manufacture." *Advanced Composites Manufacturing Centre, Plymouth University*. March 27, 2015. http://www.tech.plym.ac.uk/sme/mats324/mats324A9%20NFETE.htm (accessed April 25, 2015).
"Concern over Recycled Polyester." *The Textile Dyer*. May 13, 2008.
Corbman, Bernanrd P. 1975. *Textiles: Fibre to Fabric*. New York: McGraw Hill.
*Fabric Link, Polyolefin*. n.d. http://www.fabriclink.com/university/polyolefin.cfm (accessed April 22, 2015).
Freinkel, Susan. 2011. *Plastic: A Toxic Love Story*. New York: Houghton Mifflin Harcourt.
Hegde, Raghavendra , Atul Dahiya, and M.G. Kamath. 2004. "Nylon Fibers." *Nonwovens Science and Technology II*. April. http://www.engr.utk.edu/~mse/Textiles/Nylon%20fibers.htm (accessed June 3, 2015).
"High Density Polyethylene." 2008. *Plastics Europe*. November. http://www.plasticseurope.org/Documents/Document/20100312112214-FINAL_HDPE_280409-20081215-017-EN-v1.pdf (accessed April 25, 2015).
Marshall, Jennifer. 2010. *Discovery, Bioplastics Not so Green*. December 6. http://news.discovery.com/earth/plants/bioplastic-plant-plastic-environment.htm (accessed April 25, 2015).
"Oeko-Tex Standard 100." n.d. *Oeko-Tex*. https://www.oeko-tex.com/media/downloads/Factsheet_OETS_100_EN.pdf (accessed April 5, 2015).
"Organic Chemical Process Industry." 1991. *U.S. Environmental Protection Agency*. http://www.epa.gov/ttnchie1/ap42/ch06 (accessed April 25, 2015).

"Plastics." n.d. *U.S. Environmental Protection Agency*. http://www.epa.gov/climatechange/wycd/waste/downloads/plastics-chapter10-28-10.pdf (accessed April 25, 2015).

*PR Web*. February 14, 2012. http://www.prweb.com/releases/2012/2/prweb9194258.htm (accessed April 25, 2015).

"The Fiber Year 2009–2010." 2010. *Oerlikon Textile*. May. http://www.indotextiles.com/download/Fiber%20Year%202009_10.pdf (accessed April 21, 2015).

## Chapter 18

"Directive on Textile Names." *Lex Europa*. January 14, 2009. http://eur-lex.europa.eu/LexUriServ/LexUriServ.do?uri=OJ:L:2009:019:0029:0048:EN:PDF (accessed May 2, 2015).

*Five Bamboo*. n.d. http://fivebamboo.com/index.php/our-bamboo-fabric (accessed May 3, 2015).

*Food and Agriculture Organization of the United Nations, Future Fibres*. n.d. http://www.fao.org/economic/futurefibres/fibres/jute/en/ (accessed May 2, 2015).

"FTC Charges Companies with 'Bamboo-zling' Consumers with False Product Claims." 2009. *U.S. Federal Trade Commission*. August 11. http://ftc.gov/opa/2009/08/bamboo.shtm (accessed April 29, 2015).

*Lenzig*. n.d. http://www.lenzing.com/sites/botanicprinciples/website/index.htm.

*Litrax*. n.d. http://www.litrax.com/fibers-natural-11.html (accessed May 1, 2015).

Masters of Linen. n.d. "Be Linen Map." *Masters of Linen*. http://www.knittingindustry.com/uploads/2048/BE_LINEN_MAP_10-1.pdf (accessed March 27, 2015).

*Natural Resources Defense Council, Not All Bamboo Is Created Equal*. August 2009. http://www.nrdc.org/international/cleanbydesign/files/CBD_FiberFacts_Bamboo.pdf (accessed May 3, 2015).

*NEPCon*. October 4, 2010. http://www.nepcon.net/newsroom/dress-yourself-fsc-bamboo (Accessed May 2, 2015).

*O Ecotextiles*. n.d. http://oecotextiles.wordpress.com/2009/08/19/348/ (accessed April 29, 2015).

*O Ecotextiles*. August 19, 2009. http://oecotextiles.wordpress.com/2009/07/14/why-is-recycled-polyester-considered-a-sustainable-textile/#_ftn6 (accessed April 23, 2015).

"Process Flow Chart of Viscose Fabric Dyeing." *Textile Fashion Study*. June 13, 2012. http://textilefashionstudy.com/process-flow-chart-of-viscose-fabric-dyeing/ (accessed April 29, 2015).

*U.S. Federal Trade Commission*. August 2009. https://www.ftc.gov/tips-advice/business-center/guidance/how-avoid-bamboozling-your-customers (accessed May 3, 2015).

*Voice of America*. October 31, 2009. http://www.voanews.com/content/a-13-2006-08-29-voa51/323110.html (accessed May 2, 2015).

## Chapter 20

Gilman, Deward F., and Dennis G. Watson. *University of Florida IFAS Extension*. 2015. http://edis.ifas.ufl.edu/st244 (accessed April 17, 2015).

Lenzing. "Lenzing Modal Edelweiss." *Lenzig*. 2011. http://lenzinginnovation.lenzing.com/fileadmin/template/pdf/Texworld_USA_2012/16_01_2012_2_PM_Lenzing_Edelweiss.pdf (accessed April 17, 2015).

———. *Modal*. n.d. http://www.lenzing.com/en/fibers/lenzing-modal/applications/lenzing-modal-color.html (accessed April 17, 2015).

"Oeko-Tex Standard 100." *Oeko-Tex*. n.d. https://www.oeko-tex.com/media/downloads/Factsheet_OETS_100_EN.pdf (accessed April 17, 2015).

## Chapter 22

"Cotton Incorporated." March 2013.

Dupont. n.d. *Huntsman Gentle Power Bleach*™. http://primagreen.dupont.com/product-solutions/huntsman-gentle-power-bleachtm/.

Grose, Lynda, and Kate Fletcher. 2012. *Fashion and Sustainability: Design for Change*. London: Laurence King Publishing, Ltd (accessed May 8, 2015).

Nielson, P. H., H. Kuildred, W. Zhou, and X. Lu. 2009. "Enzyme Biotechnology for Sustainable Textiles." In *Sustainable Textiles: Lifecyle and environmental impacts*, by R. S. Blackburn, 113–138. Cambridge: Woodhead Publishing.

## Chapter 23

"AirDye Environmental Profile: Life Cycle Assessment." n.d. *AirDye Solutions*. http://www.airdyesolutions.com/uploads/AirDye_EPDv2b_091109.pdf (accessed May 21, 2015).

"Anthraquinone." n.d. *IARC Monographs*. http://monographs.iarc.fr/ENG/Monographs/vol101/mono101-001.pdf (accessed May 21, 2015).

Chequer, Farah Maria Drumond, Gisele Augusto Rodrigues de Oliveira, Elisa Raquel Anastácio Ferraz, Juliano Carvalho Cardoso, Maria Valnice Boldrin Zanzoni, and Danielle Palma de Oliveira. 2013. "Textile Dyes: Dyeing Process and Environmental Impact." Chap. 6 in *INTECH*, 151–176. doi:10.5772/53659 (accessed May 21, 2015).

Clark, James l. 1997. "Water Conservation Through Automated Dyebath Reuse." (Proceedings of the 1997 Georgia Water Resources Conference,). https://smartech.gatech.edu/bitstream/handle/1853/44194/ClarkJ-97.pdf (accessed June 2, 2015).

Dickerson, Dianne K., Eric F. Lane, and Dolores F. Rodriguez. 1999. "Naturally Colored Cotton: Resistance to Changes in Color and Durability When Refurbished With Selected Laundry Aids." Edited by California Agricultural Technology Institute. October. http://www.lsmalhas.com/colored_cotton.pdf (accessed June 2, 2015).

Kimmel, Linda, Chris Delhom, and Craig Folk. 2004. "Southern Regional Research Center Reveals Colorful New Methods." *USDA*. January 9. http://naldc.nal.usda.gov/naldc/download.xhtml?id=12180&content=PDF (accessed June 2, 2015).

Periyasamy, Aravin P., Bhaarathi Dhurai, and K. Thangamani. 2011. "Salt-Free Dyeing—A New Method of Dyeing on Cotton/Lyocell Blended Fabrics with Reactive Dyes." *AUTEX Research Journal* 11 (1): 14–17. http://www.researchgate.net/profile/Aravin_Periyasamy/publication/228842614_Salt_free_dyeingA_new_method_of_dyeing_on_LyocellCotton_blended_fabrics_with_Reactive_dyes/links/54bfbc8f0cf21674ce9c7711.pdf (accessed June 2, 2015).

Selvakumar, Sathian, Rajasimman Manivasagan, and Karthikeyan Chinnappan. 2012. "Biodegradation and decolorization of textile dye wastewater using Ganoderma lucidum." *3 Biotech* 3 (1): 71–79. doi:10.1007/s13205-012-0073-5 (accessed May 23, 2015).

## Chapter 24

Brigden, Kevin, Samantha Hetherington, Mengjiao Wang, David Santillo, and Paul Johnston. 2013. *Greenpeace Research Laboratories Techni*. December. http://www.greenpeace.org/eastasia/Global/eastasia/publications/reports/toxics/2013/A%20Little%20Story%20About%20the%20Monsters%20In%20Your%20Closet%20-%20Technical%20Report.pdf (accessed June 13, 2015).

Dow Corning. n.d. *How polydimethylsiloxane degrades in the environment*. http://www.dowcorning.com/content/discover/discoverchem/how-si-degrades.aspx (accessed June 10, 2015).

Dupont. n.d. *About PFOA*. http://www2.dupont.com/PFOA2/en_US/OandA/index.html (accessed June 13, 2015).

*European Food Safety Authority*. March 11, 2014. http://www.efsa.europa.eu/en/topics/topic/bfr.htm (accessed June 13, 2015).

Ferrero, F., and M. Periolatto. 2012. "Antimicrobial finish of textiles by chitosan UV-curing." *Journal of Nanoscience & Nanotechnology* 12 (6): 4803–4810. http://www.ncbi.nlm.nih.gov/pubmed/22905533 (accessed June 13, 2015).

Israel, Brett. 2012. "Burning irony: Flame retardants might create deadlier fires." *Environmental Health News*. April 4. http://www.environmentalhealthnews.org/ehs/news/2012/burning-irony (accessed June 13, 2015).

*Science News*. September 20, 2005. http://www.sciencedaily.com/releases/2005/09/050920002527.htm (accessed June 10, 2015).

Symptatex. n.d. *Recycleable Membrane*. http://www.sympatex.com/en/membrane/224/ecology (accessed June 10, 2015).

Tesoro, Giuliana C. 1978. "Chemical Modification of Polymers with Flame-Retardant Compounds." *Journal of Polymer Science*, 283–353. http://www.eis.uva.es/~macromol/curso07-08/ignifugos/Giulanca%20C.%20tesoro.pdf (accessed June 13, 2015).

*U.S. Environmental Protection Agency*. September 9, 2014. http://www.epa.gov/oppt/pfoa/pubs/activities.html#ord (accessed June 13, 2015).

## Chapter 25

Atav, Riza. 2013. "The Use of New Technologies in Dyeing of Proteinaceous Fibers." *InTech*. http://cdn.intechopen.com/pdfs-wm/41410.pdf (accessed May 23, 2015).

Chaudhry, G. R., and S. Chapalamadugu. 1991. "Biodegradation of halogenated organic compounds." *Microbiological Reviews* 55 (1): 59–79. http://www.ncbi.nlm.nih.gov/pmc/articles/PMC372801/pdf/microrev00032-0073.pdf (accessed May 23, 2015).

Dupont. n.d. *Huntsman Gentle Power Bleach/*. http://primagreen.dupont.com/product-solutions/huntsman-gentle-power-bleachtm/ (accessed May 8, 2015).

"Fashion victims: a report on sandblasted denim." 2010. *Fair Trade Center*. November. http://www.cleanclothes.org/resources/national-cccs/fashion-victims-a-report-on-sandblasted-denim (accessed May 23, 2015).

*Fimatex Eco-Aging*. n.d. http://www.fimatex.it/?page_id=312&lang=en (accessed May 23, 2015).

Grose, Linda, and Kate Fletcher. 2012. *Fashion and Sustainability: Design for Change*. London: Laurence King Publishing, Ltd.

Selvakumar, Sathian, Rajasimman Manivasagan, and Karthikeyan Chinnappan. 2012. "Biodegradation and decolorization of textile dye wastewater using Ganoderma lucidum." *3 Biotech*. doi:10.1007/s13205-012-0073-5 (accessed May 23, 2015).

Somerville, Heather. 2013. "Retailer sandblasting bans have changed little in the garment industry." *San Jose Mercury News*. October. http://www.mercurynews.com/business/ci_24407198/retailer-sandblasting-bans-have-changed-little-garment-industry (accessed May 23, 2015).

## Glossary

Johnston, Amanda, and Clive Hallett. 2015. *Fabric for Fashion, A comprehensive guide to natural fibers*. London: Laurence King Publishing, p. 195.

# Notes

### Front Matter

1. Ernest Callenbach, Fritjof Capra, and Sandra Marburg, Global File No 5, *Eco-Auditing and Ecologically Conscious Management* (Berkeley: Elmwood Institute, 1990).
2. Kate Fletcher and Lynda Grose, *Fashion and Sustainability: Design for Change* (London: Laurence King, 2012), 12.
3. Ernest Callenbach, *Ecology: A Pocket Guide, Revised and Expanded* (Berkeley: University of California Press, 2008), 163.
4. Local Wisdom, www.localwisdom.info, accessed June 30, 2015.

### Part 1 Overview

1. Masters of Linen 2008
2. Masters of Linen 2008
3. www.fao.org/economic/futurefibres/home/en/
4. www.hempage.de/
5. www.voanews.com/content/a-13-2006-08-29-voa51/323110.html
6. www.ncsu.edu/bioresources/BioRes_06/BioRes_06_4_5260_Paridah_ASZ_Retting_Bast_Fiber_Quality_Review_1312.pdf

### Chapter 1

1. Agriculture Marketing Resource Center, Cotton Profile 2012
2. US Department of Agriculture Economic Research Service Cotton and Wool Overview 2014
3. Bayer 2011
4. US Department of Agriculture, Adoption of Genetically Engineered Crops in the U.S. 2014
5. East 2013
6. textileexchange.org 2011
7. Organic Trade Association 2012
8. Textile Exchange, The global sustainable cotton situation n.d.
9. Bureau Veritas, Organic Cotton Frequently Asked Questions 2010
10. Sustainable Cotton Project, Better Cotton Field Program n.d.
11. Better Cotton Initiative, BCI History n.d.
12. Sustainable Cotton Project, Better Cotton Field Program n.d.
13. textileexchange.org 2011
14. Sustainable Cotton Project, Better Cotton Field Program n.d.
15. Agriculture Marketing Resource Center, Cotton Profile 2012
16. Pesticide Action Network, Cotton n.d.
17. Alter 2009
18. Glausiusz 2008
19. World Wildlife Fund, Cotton Farming 2015
20. Mirovalev and Kramer 2013
21. Bishek 2013
22. Stochaj 2009
23. www.fairtrade.net/products/cotton.html
24. Oeko-Tex® n.d.
25. textileexchange.org 2011
26. Sustainable Cotton Project, Better Cotton Field Program n.d.
27. textileexchange.org 2011
28. Sustainable Cotton Project, Cleaner Cotton 2014
29. Natural Fibres Cotton n.d.

### Chapter 2

1. Masters of Linen 2008
2. Masters of Linen 2008
3. Masters of Linen 2008
4. Masters of Linen 2008
5. *(Wattage × Hours Used Per Day) ÷ 1000 = Daily Kilowatt-hour (kWh) consumption*
   *Daily kWh consumption × number of days used per year = annual energy consumption*
6. Masters of Linen 2008
7. Mussig (Editor) 2010

### Chapter 3

1. www.voanews.com/content/a-13-2006-08-29-voa51/323110.html
2. ojs.cnr.ncsu.edu/index.php/JTATM/article/view/651/458
   Waite, M., Sustainable Textiles: The Role of Bamboo and a Comparison of Bamboo Textile Properties, JTATM Vol. 6, Issue 2, Fall 2009
3. www.voanews.com/content/a-13-2006-08-29-voa51/323110.html
4. www.voanews.com/content/a-13-2006-08-29-voa51/323110.html
5. ojs.cnr.ncsu.edu/index.php/JTATM/article/view/651/458
   Waite, M., Sustainable Textiles: The Role of Bamboo and a Comparison of Bamboo Textile Properties, JTATM Vol. 6, Issue 2, Fall 2009
6. *(Wattage × Hours Used Per Day) ÷ 1000 = Daily Kilowatt-hour (kWh) consumption*

7. business.ftc.gov/documents/alt172-how-avoid-bamboozling-your-customers
8. A quantitative antibacterial test was performed by the China Industrial Testing Centre in 2003 in which 100 percent bamboo fabric was tested with the bacterial strain type *Staphylococcus aureus*; after a twenty-four-hour incubation period, the bamboo fabric showed a 99.8 percent antibacterial destroy rate. (FAO 2007).
9. ojs.cnr.ncsu.edu/index.php/JTATM/article/view/651/458
   Waite, M., Sustainable Textiles: The Role of Bamboo and a Comparison of Bamboo Textile Properties, JTATM Vol. 6, Issue 2, Fall 2009


**Chapter 4**


1. Small and Marcus 2002
2. Small and Marcus 2002
3. Small and Marcus 2002
4. Food and Agriculture Organization of the United Nations; Future Fibres 2009
5. Small and Marcus 2002
6. Britannica n.d.
7. Britannica n.d.
8. *(Wattage × Hours Used Per Day) ÷ 1000 = Daily Kilowatt-hour (kWh) consumption*
   *Daily kWh consumption × number of days used per year = annual energy consumption*
9. Hemp Industry Association 2014
10. Small and Marcus 2002
11. Mussig (Editor) 2010
12. Small and Marcus 2002


**Chapter 5**


1. AVM Chemical n.d.
2. Mussig (Editor) 2010
3. Pure Jute n.d.
4. "Beginner's Guide to Sustainable Fibres," Textile Exchange, 2011
5. PPG Industries, Inc. 2008
6. AVM Chemical n.d.
7. New York State Department of Health 2004
8. *(Wattage × Hours Used Per Day) ÷ 1000 = Daily Kilowatt-hour (kWh) consumption*
   *Daily kWh consumption × number of days used per year = annual energy consumption*
9. Mussig (Editor) 2010
10. Food and Agriculture Organization of the United Nations n.d.
11. Mussig (Editor) 2010


**Chapter 6**


1. According to the United States Federal Trade Commission, The Wool Products Labeling Act of 1939, 15 U.S.C. § 68
2. Zhang, et al. 2014
3. www.iwto.org/uploaded/Fact_Sheets/Wool_and_Flame_Resistance_IWTO_Fact_Sheet.pdf
4. Zhang, et al. 2014
5. Russel n.d.
6. Biological Oxygen Demand (BOD) – Overview
7. Ward, 2013; Chang, 1999
8. www.nicefashion.org/en/professional-guide/recycling/Recycledtextiles.html
9. www.oeko-tex.com/media/downloads/Factsheet_OETS_100_EN.pdf


**Chapter 7**


1. Textile Exchange n.d.
2. United Nations Food and Agriculture Organization 2009
3. Slater, Environmental Impact of Textiles: Production, Processes and Protection 2003
4. Slater, Environmental Impact of Textiles: Production, Processes and Protection 2003
5. LenTech n.d.
6. McLaren 2006
7. Slater, Environmental Impact of Textiles: Production, Processes and Protection 2003
8. www.theethicalsilkco.com/eco-friendly-silk/
9. McLaren 2006
10. United Nations Food and Agriculture Organization 2009
11. United Nations Food and Agriculture Organization 2009
12. United Nations Food and Agriculture Organization 2009


**Chapter 8**


1. Mwinyihija 2010
2. Dima Nazer 2006
3. Dima Nazer 2006
4. www.osha.gov/OshDoc/data_Hurricane_Facts/hydrogen_sulfide_fact.pdf
5. Mwinyihija 2010


**Chapter 9**


1. Bruce Nelson, Personal Interview, January 5, 2014
2. Bruce Nelson, Personal Interview, January 5, 2014
3. Bruce Nelson, Personal Interview, January 5, 2014
4. Bruce Nelson, Personal Interview, January 5, 2014
5. Ziek 2012
6. Ziek 2012


**Future Fibers: Natural Fibers**


1. C. Hijosa, personal communication, June 22, 2015

2. www.crailar.com/
3. Masters of Linen 2008
4. Masters of Linen 2008

**Chapter 10**

1. The Fiber Year 2009-2010
2. Swicofil, Polyester 2013
3. New York Fashion Center, Polyester Fabric Information n.d.
4. The Fiber Year 2009-2010
5. Scaturro 2011
6. Make Wealth History, How much oil is there left, really? n.d.
7. King 2010
8. Ducas 2012
9. Ducas 2012
10. King 2010
11. Sundar and Chakravarty 2010
12. Oeko-Tex Standard 100 n.d.

**Chapter 11**

1. Freinkel 2011
2. Ensinger, Polyamides n.d.
3. Technical Bulletin 2013
4. Freinkel 2011
5. Freinkel 2011
6. Hegde, Dahiya, and Kamath 2004
7. Eur Lex 2014
8. Grose and Fletcher 2012
9. Concern over Recycled Polyester 2008
10. O Ecotextiles n.d.
11. Yamasaki n.d.
12. Upham 2009
13. Oeko-Tex Standard 100 n.d.
14. Aquafil Group Worldwide n.d.
15. Patagonia 2009

**Chapter 12**

1. Fiber Source, A Quick Guide to Manufactured Fibers n.d.
2. Fiber Source, A Quick Guide to Manufactured Fibers n.d.
3. Fiber Source, A Quick Guide to Manufactured Fibers n.d.
4. Cohen and Johnson 2010
5. Corbman 1975
6. Make Wealth History, How much oil is there left, really? n.d.
7. Adam 2010
8. Preventing Adverse Health Effects from Exposure to Dimethylformamide (DMF) 1990
9. National Service Center for Environmental Publications 2000
10. Technology Transfer Network - Air Toxics Web Site 2000
11. McKeown 2015
12. Oeko-Tex Standard 100 n.d.
13. Genomatica Sustainable Chemicals 2013
14. www.invista.com/
15. Fiber Source, A Quick Guide to Manufactured Fibers n.d.
16. Fiber Source, A Quick Guide to Manufactured Fibers n.d.
17. Fiber Source, A Quick Guide to Manufactured Fibers n.d.

**Chapter 13**

1. Cohen and Johnson 2010
2. American Chemical Society National Historic Chemical Landmarks 2007
3. Dow Chemical, Product Safety Assessment: Propylene 2006
4. Luke 2004
5. European Man-made Fibres Institute n.d.
6. Acrylonitrile Fact Sheet 1994
7. Centers for Disease Control, National Institute for Occupational Safety and Health 2015
8. Emission Factors 1990
9. Oeko-Tex Standard 100 n.d.
10. World Acrylic Fiber Report 2010–2011
11. Japan Chemical Fibers Association, Synthetic Fibers 2008
12. Oeko-Tex Standard 100 n.d.

**Chapter 14**

1. Composition Leather n.d.
2. European Commission Environment 2015
3. Healthy Child Healthy World 2013
4. Healthy Child Healthy World 2013
5. European Commission Scientific Committees Glossary n.d.
6. Phthalates entry 52 2014
7. US Consumer Product Safety Commission, Phthalates n.d.
8. US Consumer Product Safety Commission, Phthalates n.d.
9. Phthalates 2007
10. Toray Innovation by Chemistry 2013

**Chapter 15**

1. Azeem n.d.
2. Azeem n.d.
3. High Density Polyethylene 2008
4. Why Natural Fibres? 2009

**Chapter 16**

1. The Fiber Year 2009–2010 2010
2. Freinkel 2011
3. Freinkel 2011
4. Fabric Link, Polyolefin n.d.
5. Hegde, Dahiya, and Kamath 2004
6. Freinkel 2011

7. Corbman 1975
8. High Density Polyethylene 2008
9. Organic Chemical Process Industry 1991
10. Composites Design and Manufacture 2015
11. Plastics n.d.
12. PR Web 2012
13. Marshall 2010
14. Oeko-Tex Standard 100 n.d.

**Chapter 17**

1. Swicolfil, Rayon Viscose n.d.
2. Voice of America 2009
3. Canopy, Protecting Forests 2014
4. Source: Canopy. 2015. CanopyStyle: Taking the Runway by Storm. Two Years of Collaboration. www.canopyplanet.org
5. O Ecotextiles n.d.
6. Lenzig n.d.
7. Canopy, Protecting Forests 2014
8. Forest Stewardship Council n.d.
9. Lenzig n.d.

**Chapter 18**

1. Voice of America 2009
2. Voice of America 2009
3. www.hempage.de/
4. Food and Agriculture Organization of the United Nations, Future Fibres n.d.
5. Masters of Linen n.d.
6. Masters of Linen n.d.
7. Voice of America 2009
8. O Ecotextiles 2009
9. O Ecotextiles 2009
10. Lenzig n.d.
11. NEPCon 2010
12. Litrax n.d.
13. Directive on Textile Names 2009
14. Directive on Textile Names 2009
15. U.S. Federal Trade Commission 2009

**Chapter 19**

1. Schachtner 2014
2. Schachtner 2014
3. European Commission Environment, Timber 2015
4. Hao, Cai, and Fang 2009
5. How Products are Made, Lyocell
6. Lenzing, Tencel
7. Schachtner 2014

**Chapter 20**

1. Gilman and Watson 2015
2. Gilman and Watson 2015
3. Gilman and Watson 2015
4. Lenzing n.d.
5. Lenzing 2011
6. Oeko-Tex Standard 100 n.d.
7. Lenzing n.d.

**Chapter 21**

1. FiberSource, Azlon Fiber n.d.
2. www.fibre2fashion.com/industry-article/37/3699/soybean-fibers-a-review1.asp
3. Adoption of Genetically Engineered Crops in the U.S. 2014
4. Euroflax Industries, Ltd. 2005

**Future Fibers: Manufactured Fibers**

1. NatureWorks n.d.
2. Cohen and Johnson 2010
3. Henze, Renee. Fiber briefing interview with Annie Gullingsrud. San Francisco, August 10, 2016.
4. Henze, Renee. Fiber briefing interview with Annie Gullingsrud. San Francisco, August 10, 2016.

**Chapter 22**

1. Cotton Incorporated 2013
2. Nielson, et al. 2009
3. Cotton Incorporated 2013
4. Dupont n.d.
5. Cotton Incorporated 2013

**Chapter 23**

1. Chequer, et al. 2013
2. Chequer, et al. 2013
3. Chequer, et al. 2013
4. Anthraquinone n.d.
5. Mansour H., et al. 2012
6. Selvakumar, Manivasagan, and Chinnappan 2012
7. Anthraquinone n.d.
8. Chequer, et al. 2013
9. Periyasamy, Dhurai, and Thangamani 2011
10. Kimmel, Delhom, and Folk 2004
11. Dickerson, Lane, and Rodriguez 1999
12. Kimmel, Delhom, and Folk 2004
13. Kimmel, Delhom, and Folk 2004
14. Clark 1997
15. AirDye Environmental Profile: Life Cycle Assessment n.d.

**Chapter 24**

1. Science News 2005
2. U.S. Environmental Protection Agency 2014
3. Tesoro 1978
4. Dupont n.d.
5. Israel 2012
6. Tesoro 1978
7. European Food Safety Authority 2014
8. Brigden, et al. 2013

9. Dow Corning n.d.
10. Symptatex n.d.
11. Ferrero and Periolatto 2012
12. advancedtextilessource.com/2014/08/royal-dsm-announces-pfc-free-membranes-for-clothing/
13. www.greenpeace.org/eastasia/Global/eastasia/publications/reports/toxics/2013/A%20Little%20Story%20About%20the%20Monsters%20In%20Your%20Closet%20-%20Report.pdf

**Chapter 25**

1. Fashion victims: a report on sandblasted denim 2010
2. Fashion victims: a report on sandblasted denim 2010
3. Somerville 2013
4. Selvakumar, Manivasagan, and Chinnappan 2012
5. Chaudhry and Chapalamadugu 1991
6. Atav 2013
7. Fimatex Eco-Aging n.d.

**Chapter 26**

1. Mowbray 2008
2. O Ecotextiles n.d.
3. Teijin, Closed Loop Recycling System ECO CIRCLE n.d.
4. O Ecotextiles n.d.
5. O Ecotextiles n.d.
6. Mowbray 2008
7. O Ecotextiles n.d.

**Future Fibers: Promoting Circular Textiles**

1. S. Flynn, personal communication, 2014–2015
2. H. Norlin, personal communication, 2014–2015

**Appendix A**

1. Local Wisdom, www.localwisdom.info, accessed June 30, 2015.
2. Kate Fletcher, "Other Fashion Systems," in *Handbook for Sustainability and Fashion* (ed. Kate Fletcher and Matilda Tham, London, Routledge, 2014), 15–24.
3. Steven Skov Holt, conversation with author, 2012.
4. Lynda Grose and Vibeke Riisberg, unpublished studio work in process, San Francisco and Design School Kolding, 2015.

**Appendix B**

1. www.ewg.org/skindeep/ingredient/706089/SODIUM_LAURETH_SULFATE/
2. www.ewg.org/skindeep/ingredient/726331/1%2C4-DIOXANE/
3. www.epa.gov/saferchoice/partnership-evaluate-alternatives-nonylphenol-ethoxylates-publications
4. www.tfl.com/web/files/Statement_NPE-surfactants.pdf
5. europa.eu/rapid/press-release_IP-11-1542_en.htm
6. articles.mercola.com/sites/articles/archive/2011/12/21/are-you-slowly-killing-your-family-with-hidden-dioxane-in-your-laundry-detergent.aspx

**Glossary**

1. dictionary.reference.com/
2. dictionary.reference.com/
3. Better Cotton Initiative, BCI History n.d.
4. U.S. Geological Survey 2014
5. Grose and Fletcher 2012
6. Free Drinking Water n.d.
7. U.S. Environmental Protection Agency n.d.
8. McDonough and Braungart 2004
9. Chakravart, et al. 2012
10. World Health Organization 2008
11. Dissolved Oxygen and Biochemical Oxygen Demand n.d.
12. Air Pollution Emissions Overview n.d.
13. financial-dictionary.thefreedictionary.com/Fire+Resistant
14. Forest Stewardship Council n.d.
15. Global Organic Textile Standard n.d.
16. U.S. Centers for Disease Control and Prevention n.d.
17. European Environment Agency n.d.
18. Sustainable Cotton Project, Cleaner Cotton Field Program n.d.
19. International Federation of Organic Agriculture Movements n.d.
20. Concern over Recycled Polyester 2008
21. The Gold Book n.d.
22. Oeko-Tex Standard 100 n.d.
23. Programme for the Endorsement of Forest Certification n.d.
24. Perspectives Online 2002
25. What is "Sanforized"? n.d.
26. National Center for Biotechnology Information, U.S. National Library of Medicine n.d.
27. McDonough and Braungart, The Upcycle 2013
28. U.S. Environmental Protection Agency n.d.

# Index


Acid dye, 211, 269
Acrylic, 129–133
  availability, 131
  benefits, 130
  chemical treatments, 130
  chemical treatments for pilling, 130
  design, 132
  fashion applications, 131
  innovation exercises, 132, 133
  merchandising, 132
  potential impacts, 130
  potential marketing opportunities, 131
  processing, 130
  sustainable benefits, 130–131
Ahimsa silk, 74
AirDye, 221
Aldehyde-tanned leather, 81
Aliphatic solvents, 214, 269
Alpaca, 89, 90
Alpaca fibers, 89
  availability, 92
  consumer care/washing, 91–92
  design, 93
  fashion applications, 92
  innovation exercises, 93–95
  merchandising, 93, 95
  potential impacts, 91–92
  potential marketing opportunities, 93
  sustainable benefits, 90–91, 92
Angora goat, 59
Animal welfare, 269
  silk, 72
  wool, 61, 65
Antibacterial, 269
Antichlor, 269
Antimicrobials
  azlon, 189
  finishing, 226, 227, 228
Antimony free, 109, 269
APINAT, 196
Applied Separations, processing textiles, 246
Aromatic solvents, 214, 269
Australian Wool Innovation Limited (AWI), 65
Auxiliary chemicals, dyeing, 214, 218
Azlon, 103
  availability, 189
  bleaching, 188
  blends with other fibers, 188
  comparing features, 187
  cultivation, 187–188
  design, 189–190
  dyeing, 188
  fashion applications, 189
  innovation exercises, 189–190
  merchandising, 190
  potential impacts, 187–188
  potential marketing opportunities, 189
  from soy, 185–191
  sustainable benefits, 186, 188

Bamboo, 164. *See also* Bamboo linen; Rayon/viscose (from bamboo)
Bamboo linen, 29
  availability, 33
  consumer care/washing, 32
  design, 35
  dyeing, 31–32
  fashion applications, 33, 34
  marketing opportunities, 34–35
  merchandising, 35
  potential impacts, 31–33
  sustainable benefits, 30, 32
  water and chemicals use in fiber extraction, 31
Bamboo lyocell, 168
Basic dye, 211, 269
Bast fibers, 2, 3, 19, 29, 39, 49, 99
Benefits
  acrylic, 130
  finishing, 226
  imitation leather, 136
  lyocell, 172–173
  modal, 180
  nylon, 114
  polyethylene, 144
  polypropylene (PP), 150
  rayon/viscose (from bamboo), 164
  rayon/viscose (from wood), 156
  spandex, 122
Better Cotton Initiative (BCI), 8, 14, 269
Bioaccumulation/bioaccumulate, 225, 269
Bio-based, 269
  polytrimethylene terephthalate, 195
  spandex, 124, 125
Biodegradability/biodegradable, 269
  alpaca fibers, 93
  azlon, 189
  cotton, 10
  Evrnu, 260
  lyocell, 174
  nylon, 115
  polylactide (PLA), 194
  polytrimethylene terephthalate (PTT), 195
  Re:newcell, 262, 263
  silk, 75
  wool, 67
Biodiversity, *xiii*, 6, 10, 13, 269
  EcoPlanet Bamboo, 167
  wool, 62, 65, 66
Biologically efficient, 269
Biological oxygen demand (BOD), 62, 269
Bionic DPX®, 196
Bleaching, 201
  alternative technologies, 202–203, 239
  availability, 204
  azlon, 188
  chlorine derivatives, 202
  design, 206
  fashion applications, 204–205
  garment washing, 235
  hydrogen peroxide, 202
  innovation exercises, 206
  laccase, 203
  merchandising, 206
  ozone, 202–203
  potential marketing opportunities, 205
  sustainable benefits, 204
  types of, 202
Blend azlon and natural fibers, 188
Bonded leather, 137
Brightening, garment washing, 235

Cadmium, 136, 269
California Cloth Foundry, 205
Canopy, 156
Carbon neutral, 180, 269
Carcinogen, 270
Cashmere, blend with azlon (from soy), 188
Cashmere goat, 59
Cationic fixing agents, 217, 270
Cellulosic fibers, 2. *See also* Manufactured cellulosic fibers
Chemically recycled polyester, 109
Chemical oxygen demand (COD), 212, 270
Chemical processing, recycled/circular textile technologies, 253–254
Chemical recycling, 255, 270
  nylon, 115
  polyethylene terephthalate, 253–254
Chemical retting, 3
Chemical selection, garment washing, 238
Chemical treatments for pilling, acrylic, 130

Chemical use for processing, wool, 62
Chemical use in cultivation
  bamboo linen, 31
  cotton, 11
  flax, 21
Chemical use in extracting fiber
  bamboo linen, 31
  hemp, 41
  jute, 51
Chemical use in processing, leather, 80–82
Chitosan, 228, 270
Chlorine derivatives, 202, 270
Chrome tanning, leather, 81, 83
Chromium, environmental impact, 213–214
Chromophoric, 215, 270
Chronic, 270
Circular textiles, 249, 259
Cleaner Cotton™, 8–9, 12, 14
Climate change, *xiii*, 270
Closed-loop process, 270
  lyocell fiber, 171, 172, 177
  nylon, 115
  rayon/viscose (from bamboo), 166–167
  rayon/viscose (from wood), 155, 158
$CO_2$ assimilation rate, jute, 51
ColorZen, processing textiles, 246
Compostable, 144, 145, 151, 270
Consumer care
  alpaca fibers, 91
  bamboo linen, 32
  cotton, 11
  environmental impacts, 267
  flax, 22
  hemp, 41
  jute, 53
  leather, 82
  nylon, 114–115
  rayon/viscose (from bamboo), 165
  rayon/viscose (from wood), 156–157
  silk, 74
  wool, 63
Cotton, 2, 5
  availability, 12, 14
  blend with azlon (from soy), 188
  certification, 12
  certified organic, 6–7
  chemical use in cultivation, 11
  colored, 9–10
  consumer care/washing, 11
  conventional, 6
  design, 15
  fashion applications, 14
  forced labor, 11
  genetically modified, 6
  merchandising, 15, 17
  naturally colored, 217–218
  organic in-conversion, 5, 7, 8, 15, 272
  potential impacts, 11–12
  potential marketing opportunities, 14–15
  sustainable benefits, 10, 13
  transitional organic cotton, 7
  water use in cultivation, 11
CRAiLAR flax fiber, 98–99, 101, 270
Cross-linking agent, 214
Cultivation
  azlon, 187–188
  rayon/viscose (from bamboo), 168
  rayon/viscose (from wood), 156
Cultural sustainability, 265

Decompose, 270
Deforestation, xiii, 270
Degradability, 270
Denim, garment washing process, 234–236
Desertification, 66, 90, 91, 95, 270
Design
  acrylic, 132
  alpaca, 93
  azlon, 189–190
  bamboo linen, 35
  bleaching, 206
  cotton, 15
  dyeing, 222
  finishing, 229
  flax, 25
  garment washing, 241–242
  hemp, 44, 46
  imitation leather, 139
  jute, 55
  leather, 85
  lyocell, 176–177
  modal, 182
  nylon, 118
  Paper No. 9, 101
  polyester, 109
  polyethylene, 146
  polypropylene, 153
  polytrimethylene terephthalate (PTT), 197
  printing, 222
  rayon/viscose (from bamboo), 168
  rayon/viscose (from wood), 160
  recycled/circular textiles technologies, 256
  Re:newcell, 263
  silk, 75
  spandex, 126
  wool, 67
Desizing, garment washing, 234
Dew retted, 35, 37, 44, 270
Dew retting, 3, 21, 41, 42
Digital printing, 219
Dimethylformamide (DMF), 123
Direct printing, 215, 270
Discharge printing
  environmental impact, 216
  removal of color, 215–216
  thiourea dioxide, 216
  zinc formaldehyde-sulphoxylate, 216
Disperse dye, 211, 270
Dissolved oxygen (DO), 62, 270
Dorlastan® (Bayer), 121
Downcycling, 64, 249, 270
Dry-cleaning, 267
Dry garment washing, 233
Durable water repellant (DWR), 114, 270
Dyebath reuse, 218–219
Dyeing, 209–210
  AirDye, 221
  auxiliary chemicals, 218
  azlon, 188
  bamboo linen, 31–32
  cotton, 10, 12
  design, 222
  dyebath reuse, 218–219
  environmental concerns, 210
  flax, 22
  hemp, 41
  innovation exercises, 222, 223
  innovation possibilities, 220–221
  jute, 51, 53
  low-liquor-ratio, 218
  merchandising, 222
  modal, 180–181
  natural colorants, 212–214
  pad-batch, 219
  potential marketing opportunities, 221
  reactive and direct dyes, 216–217
  reducing environmental impacts, 217
  "right-first-time," 218
  sustainability, 220–221
  synthetic colorants, 211–212
  types of, 211
  wool, 62–63
DyStar® Indigo Vat dye, 247

Eco-aging, 240, 270
ECO CIRCLE™, 253, 254
ECONYL® Regeneration System, 196
EcoPlanet Bamboo, 167
Elastane. *See* Spandex
Embodied energy and resources, polyester, 106
Emissions, *xiii*, 270
  $CO_2$, *xiii*, 65, 144, 151, 255, 259, 260, 262
  greenhouse gas, 6, 10, 11, 13, 194
  hydrogen sulfide, 80, 81
  sulfur dioxide, 81
Emulsion inks, 214, 270
End of use, polyethylene, 144
Environmental impact
  discharge printing, 216
  dyeing, 210, 217
  natural colorants, 213–214
  printing, 210
  reducing, of dyeing and printing, 217–221

Enzyme retted, 25, 27, 35, 37, 44, 47, 55, 57, 270
Enzymes, 270–271
Evrnu, 260

Fabric selection, garment washing, 237
Fair Trade, 12, 15, 74, 271
Fashion industry, sustainability and, *xiv*
Fast fashion, 105, 271
Fast-growing natural resource, 271
Federal Trade Commission (FTC), 189, 273
  bamboo linen, 34
  viscose from bamboo, 168
Finishing, 225–231
  availability, 229
  benefits, 226
  fashion applications, 229
  innovation exercises, 229–230, 231
  potential impacts, 226–227
  potential marketing opportunities, 229
  sustainable impacts, 227–228
Fire resistant, 60, 94, 271
Flame retardants, finishing, 226, 227, 228
Flax, 19
  availability, 22
  consumer care/washing, 22
  CRAiLAR flax fiber, 98–99
  design, 25
  dyeing, 22
  fashion applications, 24
  fast-growing renewable fibers, 164
  marketing opportunities, 24–25
  merchandising, 25, 27
  potential impacts, 21–22
  sustainable benefits, 23
Fly-strike, sheep, 61
Forest Stewardship Council (FSC), 157, 160, 166, 167, 168, 172, 174, 182, 271
Future fibers
  Apinat Bioplastics, 196
  Bionic DPX®, 196
  CRAiLAR flax fiber, 98–99
  ECONYL, 196
  Evrnu, 260
  innovation exercises, 101
  Paper No. 9, 99–100
  Pinatex™, 97–98
  polylactide (PLA), 193–194
  polytrimethylene terephthalate (PTT), 194–195, 197
  processing, 245–247
  promoting circular textiles, 259
  Re:newcell, 261, 262–264

Garment, stages of life cycle, *xiii*
Garment washing, 233–234. *See also* Washing
  bleaching alternatives, 239
  combination/elimination of, 239–240
  design, 241–242
  eco-aging, 240
  equipment types, 234
  fabric selection, 237
  innovation exercises, 241–242, 243
  lasers, 239
  low-liquor-ratio washing, 238
  machine cleanings, 238
  merchandising, 242
  potential impacts, 236–237
  potential marketing opportunities, 241
  proper chemical selection, 238
  source reduction, 240
  steps for typical denim process, 234–236
  sustainable possibilities of, 241
  techniques for minimizing issues, 237–240
  waste minimization, 240
  water reuse, 237–238
  wet abrasion, 235, 243
Genetically modified cotton, 6, 271
Global organic cotton production, 7
Global Organic Textile Standard (GOTS), 12, 15, 23, 42, 52, 55, 271
Glutaraldehyde, 81, 271
Greenpeace, 227, 229

Halogenated organic compounds (AOX), 22, 31, 41, 51, 188, 237, 271
Hemp, 39
  availability, 43
  consumer care/washing, 41
  design, 44, 46
  dyeing, 41
  fashion applications, 43–44
  fast-growing renewable fibers, 164
  marketing opportunities, 44
  merchandising, 46
  potential impacts, 41
  rapidly renewable, 40
  sustainable benefits, 40, 42
  undyed fiber, 40
  water and chemicals used for extracting fiber, 41
Herbicide tolerant (Ht) GM cotton, 6, 10, 271
Human carcinogens
  flax dyeing, 22
  hemp dyeing, 41
Humectant, 216, 271
Hydrogen disulfide, 156, 164, 271
Hydrogen peroxide, 202, 271
Hydrogen sulfide, leather, 80
Hypochlorite, 202

Imitation leather, 135–141. *See also* Leather
  availability, 137
  benefits, 136
  design, 139
  fashion applications, 137–138
  innovation exercises, 139–141
  merchandising, 141
  potential impacts, 136
  potential marketing opportunities, 138–139
  substitutes for 100 percent genuine leather, 137
  sustainable benefits, 137, 138
  vegan leather, 138, 139
Indirect printing, 215, 271
Ingeo, NatureWorks LLC, 193
Ink-jet printing, 219
Innovation exercises
  acrylic, 132, 133
  alpaca fibers, 93–95
  azlon, 189–190
  bamboo linen, 35–36
  cotton, 15–17
  dyeing, 222, 223
  finishing, 229–230, 231
  flax, 25–27
  future fibers, 101
  garment washing, 241–242, 243
  hemp, 44–46
  imitation leather, 139–141
  jute, 55–56
  leather, 85–86
  modal, 182, 183
  nylon, 118–119
  polyester, 109–111
  polyethylene, 146
  printing, 222
  rayon/viscose (from bamboo), 168–169
  rayon/viscose (from wood), 160
  recycled/circular textiles technologies, 256–257
  Re:newcell, 263–264
  silk, 75–77
  spandex, 126
  wool, 67–69
Innovation opportunities
  dyeing and printing, 220–221
  lyocell, 176–177
  polypropylene (PP), 153
  rayon/viscose (from bamboo), 168–169
Insoluble, disperse dyes, 211
Integrated Pest Management (IPM), 8–10, 271
International Federation of Agriculture Movements (IFOAM), 271
  bamboo linen, 32
  cotton, 7
  flax, 20, 23
  hemp, 40, 42
  jute, 51, 52
  silk, 73
INVISTA, bio-based spandex, 124, 125
Ioniqa, circular technologies, 262

Ironing, 267
Irrigation, 11, 271

Jeanologia, processing textiles, 247
Jute, 49
  availability, 53
  biologically efficient, 50
  consumer care/washing, 53
  design, 55
  dyeing, 51, 53
  fashion applications, 53, 54
  fast-growing renewable fibers, 164
  merchandising, 55
  potential impacts, 51, 53
  potential marketing opportunities, 54–55
  sustainable benefits, 50–51, 52
  water and chemicals used for extracting fiber, 51

Laccase
  bleaching, 203
  garment washing, 239
Land use, wool, 62
Lasers, garment washing, 239
Laundry detergent, 267
Lead, 136, 271
Leather, 79. *See also* Imitation leather
  alternative to traditional chrome tanning, 84
  availability, 83
  chemical use in processing, 80–82
  comparing tanning processes for, 81
  consumer care/washing, 82
  design, 85
  fashion applications, 84
  innovation exercises, 85–86
  merchandising, 85
  potential impacts, 80–82
  recyclability, 82
  sustainable benefits, 80, 83
Lenzing
  Modal®, 179, 180
  TENCEL® lyocell, 171, 180
  Viscose®, 180
Linen, flax plant, 19
Low-liquor-ratio
  dyeing, 218
  garment washing, 238
Low water footprint in cultivation, 24, 34, 44, 54, 168, 271
LYCRA® (DuPont), 121
Lyocell, 171, 271
  availability, 174
  benefits, 172–173
  closed-loop system, 171, 172, 177
  design, 176–177
  dyeing process, 173
  fashion applications, 174
  innovation opportunities, 176–177
  merchandising, 177
  potential impacts, 173
  potential marketing opportunities, 174
  process, 158, 160, 166, 172
  sustainable benefits, 174
  TENCEL®, 173, 174, 175, 176, 177, 182, 190

McCartney, Stella, 138, 139
Machine cleanings, garment washing, 238
Manufactured cellulosic fibers, 103, 271
  rayon/viscose (from wood), 155–161
Manufactured fibers
  acrylic, 129–133
  nylon, 113–119
  overview, 103
  polyester, 105–111
  polyethylene, 143–147
  polylactide (PLA), 193–194
  polypropylene, 149–153
  polytrimethylene terephthalate (PTT), 194–197
  spandex, 121–127
Manufactured synthetic fibers, 103, 271
  acrylic, 129–133
  imitation leather, 135–141
  nylon, 113–119
  polyester, 105–111
  polyethylene, 143–147
  polypropylene, 149–153
  spandex, 121–127
Marketing opportunities
  alpaca fibers, 93
  azlon, 189
  bamboo linen, 34–35
  bleaching, 205
  cotton, 14–15
  dyeing, 221
  finishing, 229
  flax, 24–25
  garment washing, 241
  hemp, 44
  imitation leather, 138–139
  jute, 54–55
  lyocell, 174, 175, 176
  modal, 182
  nylon, 117–118
  polyester, 109
  polyethylene, 146
  polypropylene (PP), 152
  polytrimethylene terephthalate (PTT), 195
  printing, 221
  rayon/viscose (from bamboo), 168
  rayon/viscose (from wood), 160
  recycled/circular textiles technologies, 256
  Re:newcell, 263
  silk, 74–75
  spandex, 125
  wool, 67
Mechanical recycling, 252, 255, 271–272
  hemp, 40
  nylon, 115
  wool, 63
Melt processing, manufactured fibers, 252–253, 255
Merchandising
  acrylic, 132
  alpaca, 93, 95
  azlon, 190
  bamboo linen, 35
  bleaching, 206
  cotton, 15, 17
  dyeing, 222
  finishing, 230
  flax, 25, 27
  garment washing, 242
  hemp, 46
  imitation leather, 141
  jute, 55
  leather, 85
  lyocell, 177
  modal, 182
  nylon, 118–119
  Paper No. 9, 101
  polyester, 111
  polyethylene, 146
  polypropylene, 153
  polytrimethylene terephthalate, 197
  printing, 222
  rayon/viscose (from bamboo), 169
  rayon/viscose (from wood), 160
  recycled/circular textiles technologies, 256–257
  Re:newcell, 263–264
  silk, 77
  spandex, 126
  wool, 67, 69
Merino sheep, wool, 59, 60, 61
Modal, 179–183
  availability, 181
  benefits, 180
  design, 182
  dyeing, 180–181
  fashion applications, 181
  finishing, 180–181
  innovation exercises, 182, 183
  merchandising, 182
  potential impacts, 180–181
  potential marketing opportunities, 182
  sustainable benefits, 181
Monomer, 150, 272
Mordant, 62, 272
Motes, 203, 272
Mutagens, 272
  flax dyeing, 22
  hemp dyeing, 41

Naphthol dye, 211, 272
Natural colorants
  dye colors, 213

dyeing, 212–214, 223
environmental impacts, 213–214
Natural fibers, 2
"Naturally tanned" leather, 85
Nonchloride bleached, 205
Non-mulesed wool, 66, 67
Nonrenewable resource, 114, 272
Novozymes, processing textiles, 247
Nylon (6 and 6,6), 113–119
availability, 117
benefits, 114
biodegradability, 115
consumer care/washing, 114–115
design, 118
fashion applications, 117
innovation exercises, 118–119
merchandising, 118–119
potential impacts, 114–115
potential marketing opportunities, 117–118
processing, 114
recyclability, 115
sustainable benefits, 116

OE 100 Standard, 272
OEKO-TEX® certified, 272
acrylic, 131
cotton, 12, 15
modal, 181, 182
nylon, 116, 117, 118
polypropylene, 151
rayon, 159, 160
spandex, 123–124
wool, 66
Organic, 189, 272
Organic Exchange in Europe, 7
Organic in-conversion cotton, 5, 7, 8, 15, 272
Organic leather, 85
Organic silk, 74, 75
Organochlorines, garment washing, 237
Organotins, 136, 227, 228, 230
Overdyeing, garment washing, 235–236
Oxygenated solvents, 214, 272
Ozone, 272
bleaching, 202–203
garment washing, 239, 240

Pad-bath dyeing, 219
Paper No. 9, 99–100
availability, 100
design, 101
fashion applications, 100
merchandising, 101
potential impacts, 100
sustainable benefits, 99–100
Perfluorooctane sulfonate (PFOS), 114, 116
Perfluorooctanoic acid (PFOA), 114, 116, 227–228
Pinatex™, 97–98, 101
Plastisol inks, 214, 215, 272
Pollution prevention, garment washing, 239–240
Polyester, 105
availability, 108
benefits, 106
design, 109
fashion applications, 109
innovation exercises, 109–111
merchandising, 111
potential impacts, 106–107
potential marketing opportunities, 109
processing, 106–107
sustainable benefits, 107, 108
Polyethylene (PE), 273
applications, 145
availability, 145
benefits, 144
design, 146
end of use, 144
innovation exercises, 146
merchandising, 146
potential impacts, 144
potential marketing opportunities, 146
processing, 144
sustainable benefits, 144
types of, 144
Polyethylene terephthalate (PET), 105
chemical recycling, 253–254
mechanical recycling, 194, 256, 271–272
melt processing, 252–253
Polylactide, 272
availability, 194
biodegradability, 194
fashion applications, 194
recyclability, 194
sustainable benefits, 194
Polypropylene (PP), 149–153
availability, 152
benefits, 150
design, 153
fashion applications, 152
innovation opportunities, 153
merchandising, 153
potential impacts, 150–151
potential marketing opportunities, 152
processing, 150–151
sustainable benefits, 151
Polytrimethylene terephthalate (PTT), 194–195, 197
availability, 195
biodegradability, 195
design, 197
fashion applications, 195
merchandising, 197
potential impacts, 195
potential marketing opportunities, 195
recyclability, 195
sustainable benefits, 194, 195
Polyurethane laminate (PUL), 135, 136, 272
Polyvinyl chloride (PVC), 135, 136, 215
Porous, 30, 37, 81, 273
Post-consumer recycled content, 272
cotton, 15
leather, 138
wool, 67
Post-consumer waste, 272
nylon, 115
recycled fibers, 251
Post-industrial waste, 272
cotton, 14
manufacturing, 251
nylon, 115
Potential impacts
acrylic, 130
bamboo linen, 31–33
bleaching, 202
cotton, 11–12
Evrnu, 260
finishing, 226–227
flax, 21–22
garment washing, 236–237
hemp, 41
imitation leather, 136
jute, 51, 53
leather, 80–82
lyocell, 173
nylon, 114–115
Paper No. 9, 100
polyethylene, 144
polypropylene (PP), 150–151
polytrimethylene terephthalate (PTT), 195
rayon/viscose (from bamboo), 164–165
rayon/viscose (from wood), 156–157
Re:newcell, 262
silk, 72–74
spandex, 122–123
synthetic colorants, 212
wool, 61–63
Printing, 214. *See also* Dyeing
design, 222
digital, 219
discharge (removal of color), 215–216
dyeing and, 209–210
emulsion inks, 214
environmental concerns, 210
ink-jet, 219
innovation exercises, 222
innovation possibilities, 220–221
merchandising, 222
plastisol inks, 214, 215
potential marketing opportunities, 221
reducing environmental impacts, 217
sustainability, 220–221
Processing, 199
acrylic, 130
future fibers, 245–247

Processing *(continued)*
- innovative alternative methods, 246
- lyocell, 172
- modal, 180–181
- nylon, 114
- overview, 199
- polyester, 106–107
- polyethylene, 144
- polypropylene, 150–151
- rayon/viscose (from bamboo), 164–165
- rayon/viscose (from wood), 156, 157
- spandex, 122–123

Product life cycle, *xiii*
Programme for the Endorsement of Forest Certification (PEFC), 166, 167, 172, 174, 182, 272
Propylene, 272–273
Protein fiber, 103, 273
Protein fibers, 2
PTT. *See* Polytrimethylene terephthalate (PTT)
PU coated split leather, 137, 273

Rapidly renewable, 40, 273
Rayon/viscose (from bamboo), 163–169
- availability, 167
- benefits, 164
- closed-loop system, 166–167
- consumer care/washing, 165
- design, 168
- fashion applications, 167
- fast-growing renewable fibers, 164
- innovation opportunities, 168–169
- merchandising, 169
- potential impacts, 164–165
- potential marketing opportunities, 168
- processing, 164–165
- sustainable benefits, 166–167

Rayon/viscose (from wood), 155–161
- availability, 158–159
- benefits, 156
- closed-loop system, 155, 158
- consumer care/washing, 156–157
- cultivation, 156
- design, 160
- fashion applications, 159
- innovation exercises, 160
- merchandising, 160
- potential impacts, 156–157
- potential marketing opportunities, 160
- processing, 156, 157
- sustainable benefits, 157–158

Reactive dye, 211, 273
Recyclability/recyclable, 273
- Evrnu, 260
- nylon, 115, 118
- polylactide, 194
- polytrimethylene terephthalate, 195
- wool, 63–64

Recycled/circular textile technologies, 251–252
- chemical recycling, 253–254
- comparison chart, 255
- design, 256
- innovation exercises, 256–257
- mechanical recycling, 252
- melt processing, 252–253
- merchandising, 256–257
- potential marketing opportunities, 256
- sustainable benefits, 254–255

Recycled content
- nylon (x-percent), 117
- Polyester (100 percent), 109

Recycled fibers, 249, 251, 262
- chemical recycling, 254
- lyocell, 177
- mechanical recycling, 252, 255
- rayon/viscose, 161, 169
- regenerated fibers, 259, 262
- from textile industrial waste, 256
- wool, 61

Recycling
- chemical, 253–254
- mechanical, 252

Renewable fibers, 3, 164
Renewable natural resource, 74, 189, 273
Renewable resource, alpaca fibers, 93
Re:newcell, 261, 262–264
- availability, 263
- biodegradability/recyclability, 262
- design, 263
- fashion applications, 263
- innovation exercises, 263–264
- merchandising, 263–264
- potential impacts, 262
- potential marketing opportunities, 263
- sustainable benefits, 262, 263

Retting, 21, 31, 273
Retting process comparison chart, 3
Reuse, wool, 63
"Right-first-time" dyeing, 218
Roundup Ready, 6, 273

Sanforizing, 234, 240, 273
Scouring, 273
- garment washing, 234
- wool, 62

Sericin, 73, 273
Sheep, wool, 59
Shrink-proof treatments, wool, 62
Silk, 71
- animal welfare, 72
- availability, 74
- blend with azlon (from soy), 188
- consumer care/washing, 74
- design, 75
- fashion applications, 74
- innovation exercises, 75–77
- merchandising, 77
- potential impacts, 72–74
- potential marketing opportunities, 74–75
- processing, 73
- promoting use of organic, 74
- sizing agent, 73
- sustainable benefits, 72, 73

Silkworms, 71
Sizing agent, 273
Social sustainability, 265
Softening, garment washing, 236
Sorona®. *See* Polytrimethylene terephthalate (PTT)
Source reduction, garment washing, 240
Spandex, 121
- availability, 124
- benefits, 122
- design, 126
- fashion applications, 124, 125
- innovation exercises, 126
- merchandising, 126
- potential impacts, 122–123
- potential marketing opportunities, 125
- processing, 122–123
- sustainable benefits, 123–124
- swimsuit, 125

Stain repellent, 273
- alpaca fibers, 90
- finishing, 226, 227, 228

Staple, 273
Staple length, 10
Strike off, 219, 273
Sulfur dye, 211, 273
Surfactant, 211–212
Sustainability
- cultural, 265
- fashion industry and, *xiv*
- fibers, 2
- social, 265

Sustainable benefits
- acrylic, 130–131
- alpaca fibers, 90–91, 92
- azlon, 186, 188
- bamboo linen, 30, 32
- bleaching, 204
- cotton, 10, 13
- dyeing, 220–221
- Evrnu, 260
- flax, 20, 23
- garment washing, 241
- hemp, 40, 42
- imitation leather, 137, 138
- jute, 50–51, 52
- leather, 80, 83
- lyocell, 174
- modal, 181
- nylon, 116
- Paper No. 9, 99–100
- polyester, 107, 108

polyethylene, 144
polylactide, 194
polypropylene (PP), 151
polytrimethylene terephthalate (PTT), 195
printing, 220–221
rayon/viscose (from bamboo), 166–167
rayon/viscose (from wood), 157–158
recycled/circular textiles technologies, 254–255
Re:newcell, 263
silk, 72, 73
spandex, 123–124
wool, 60, 65–66
Sustainable Cotton Project (SCP) Cleaner Cotton, 8–9, 273
Synthetic colorants, dyeing, 211–212
Synthetic cotton, 6
Synthetic fibers. *See* Manufactured synthetic fibers

Tannin, 81, 273
Tanning, leather, 80–82
TENCEL® lyocell, 173, 174, 175, 176, 177, 182, 190
Teratogens, 273
flax dyeing, 22
hemp dyeing, 41
Textile Exchange, 65, 66
Thermoplastic polyurethane (TPU), 135, 136
Thermoplastics, 273
Thiourea dioxide, 216
Tinting, garment washing, 235–236
2,4-Toluene diisocyanate (TDI), 123, 136
Transitional cotton, 7, 273
Tributyltin (TBT), 227

Undyed, alpaca fibers, 93
United States Environmental Protection Agency (EPA), 130, 273
United States Federal Trade Commission (FTC). *See* Federal Trade Commission (FTC)
Upcycling, 64, 249, 273
U.S. Department of Agriculture (USDA), 7, 9

Vat dye, 211, 247, 273–274
Vegan leather, 137, 138, 139, 141
Vegetable tanning, leather, 81, 83, 85
Viscose, 168. *See also* Rayon/viscose (from bamboo); Rayon/viscose (from wood)
Volatile organic compounds (VOCs), 130, 274
Volatility, 218, 274

Washing
alpaca fibers, 91
bamboo linen, 32
cotton, 11
dry-cleaning, 267
environmental impacts, 267
flax, 22
hemp, 41
ironing, 267
jute, 53
laundry detergent, 267
leather, 82
nylon, 114–115
rayon/viscose (from bamboo), 165
rayon/viscose (from wood), 156–157
silk, 74
wool, 63
Waste minimization, garment washing, 240
Water conservation, bleaching, 205
Water repellent, 90, 274
finishing, 226–228
Water retting, 3
Water reuse, garment washing, 237–238
Water use in cultivation
bamboo linen, 31
cotton, 11
flax, 21
Water use in extracting fiber
bamboo linen, 31
hemp, 41
jute, 51
Wet abrasion, garment washing, 235, 243
Wet garment washing, 233
Wet-green® tanning agent, leather, 83, 84
Wild silk, 74
Wood. *See* Rayon/viscose (from wood)
Wool, 59
animal welfare, 61
availability, 66
biodegradable, 67
blend with azlon (from soy), 188
chemical use for processing, 62
consumer care/washing, 63
design, 67
dyeing, 62–63
fashion applications, 66
fiber characteristics, 60
innovation exercises, 67–69
land use, 62
merchandising, 67, 69
non-mulesed, 66, 67
potential impacts, 61–63
potential marketing opportunities, 67
recyclability, 63–64
reuse, 63
shrink-proofing treatments, 62
specialty fibers, 59
sustainable benefits, 60, 65–66
Worn Again, circular technologies, 262

Zinc formaldehyde-sulphoxylate, 216